Data Mining - Concepts and Applications

Edited by Ciza Thomas

Published in London, United Kingdom

IntechOpen

Supporting open minds since 2005

Data Mining - Concepts and Applications
http://dx.doi.org/10.5772/intechopen.95167
Edited by Ciza Thomas

Contributors
Yoosoo Oh, Seonghee Min, Andri Irfan Rifai, Setiawan Hadi, Paquita Putri Ramadhani, Julius Olufemi Ogunleye, Yao Shan, Esma Ergüner Özkoç, P.V. Sai Charan, P. Mohan Anand, Sandeep K. Shukla, Mawande Sikibi, Wencai Du, Weijun Li, Qun Yang, Leon Bobrowski, Farzaneh Mansoori Mooseloo, Saeid Sadeghi, Maghsoud Amiri, Wei-Cheng Ye, Jia-Ching Wang

First published in London, United Kingdom, 2022 by IntechOpen
IntechOpen is the global imprint of INTECHOPEN LIMITED, registered in England and Wales, registration number: 11086078, 5 Princes Gate Court, London, SW7 2QJ, United Kingdom
Printed in Croatia

British Library Cataloguing-in-Publication Data
A catalogue record for this book is available from the British Library

Additional hard and PDF copies can be obtained from orders@intechopen.com

Data Mining - Concepts and Applications
Edited by Ciza Thomas
p. cm.

This title is part of the Artificial Intelligence Book Series, Volume 8
Topic: Machine Learning and Data Mining
Series Editor: Andries Engelbrecht
Topic Editor: Marco Antonio Aceves Fernandez

Print ISBN 978-1-83969-266-6
Online ISBN 978-1-83969-267-3
eBook (PDF) ISBN 978-1-83969-268-0
ISSN 2633-1403

For EU product safety concerns:
IN TECH d.o.o., Prolaz Marije Krucifikse Kozulić 3, 51000 Rijeka, Croatia, info@intechopen.com or visit our website at intechopen.com.

IntechOpen Book Series

Artificial Intelligence

Volume 8

Aims and Scope of the Series

Artificial Intelligence (AI) is a rapidly developing multidisciplinary research area that aims to solve increasingly complex problems. In today's highly integrated world, AI promises to become a robust and powerful means for obtaining solutions to previously unsolvable problems. This Series is intended for researchers and students alike interested in this fascinating field and its many applications.

Meet the Series Editor

Andries Engelbrecht received the Masters and Ph.D. degrees in Computer Science from the University of Stellenbosch, South Africa, in 1994 and 1999 respectively. He is currently appointed as the Voigt Chair in Data Science in the Department of Industrial Engineering, with a joint appointment as Professor in the Computer Science Division, Stellenbosch University. Prior to his appointment at Stellenbosch University, he has been at the University of Pretoria, Department of Computer Science (1998-2018), where he was appointed as South Africa Research Chair in Artifical Intelligence (2007-2018), the head of the Department of Computer Science (2008-2017), and Director of the Institute for Big Data and Data Science (2017-2018). In addition to a number of research articles, he has written two books, Computational Intelligence: An Introduction and Fundamentals of Computational Swarm Intelligence.

Meet the Volume Editor

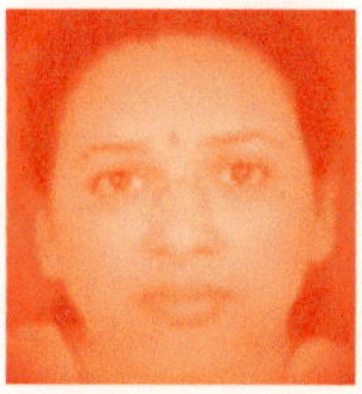

Dr. Ciza Thomas is currently a Senior Joint Director at the Directorate of Technical Education, Government of Kerala, India. Her area of expertise is network security with research interest in the fields of information security, data mining, sensor fusion, pattern recognition, information retrieval, digital signal processing, and image processing. She has more than eighty journal papers and fifty conference publications to her credit. She has edited nine books and published sixteen book chapters. She is a reviewer of more than ten international journals including *IEEE Transactions on Signal Processing, IEEE Transactions on Neural Networks, International Journal of Network Security, International Journal of Network Management*, and *Security and Communications Network*. Dr. Thomas received an achievement award in 2010 and an e-learning IT award in 2014 from the Government of Kerala.

Contents

Preface

The digitization and subsequent automation of global activities have considerably enhanced our capabilities for both creating and amassing data from various sources. This has resulted in a great amount of data flood in almost every facet of our lives. The explosive growth in warehoused data has generated an urgent need for new techniques and automated tools that can logically support us in converting available big data into useful information and knowledge. Data mining is a promising, leading-edge technology for mining large volumes of data for knowledge discovery. Data mining algorithms can be used either for clear description of data or for prediction of future outcomes from data. This can be accomplished through characterization, summarization, association, clustering, classification, discrimination, anomaly detection, trend or evolution prediction, and much more. Accordingly, data mining can be either descriptive or predictive.

Researchers and practitioners in statistics, pattern recognition, machine learning, artificial intelligence, data analytics, and visualization are contributing to the field of data mining for better utilization of data. Data mining finds applications in the entire spectrum of science and technology including basic sciences to life sciences and medicine, to social, economic, and cognitive sciences, to engineering and computers. Data mining also finds tremendous applications in business analytics.

This book discusses the concepts of data mining and presents some of the advanced research in this field. The book provides the fundamentals, techniques, and methods of processing big data for various applications. The chapters discuss the concepts, applications, and research frontiers in data mining with algorithms and implementation details for use in the real world. It includes twelve chapters divided into two sections: "Concepts of Data Mining" and "Applications of Data Mining." The initial seven chapters describe the concepts of data mining, while the remaining five chapters discuss the applications of data mining. The chapters include real-world problems in various fields and propose methods to address them. The first chapter introduces readers to the technologies explored in each of the subsequent chapters.

Chapter 1 provides an overview of the data mining process and its benefits and drawbacks, as well as discusses data mining methodologies and tasks. This chapter also discusses data mining techniques in terms of their features, benefits, drawbacks, and application areas.

After the introductory chapter on the concepts of data mining, we look at the various steps in the data mining process. The initial step after acquiring the data is data cleaning, followed by data Integration, data reduction, and data transformation. The data is then analyzed and evaluated for knowledge discovery.

Chapter 2 describes the initial step of data cleaning to prepare data for strategic decisions. As the pre-processing of data is an important step in the data mining process, the data cleaning process helps in obtaining accurate strategic decisions. The presence of incorrect or inconsistent data can significantly distort the results of analyses, often negating the potential benefits of strategic decision-making approaches. Thus, the representation and quality of data are first and foremost before running an analysis. As such, this chapter identifies the sources of data collection to remove errors and describes data mining cleaning and its methods.

Privacy has become a serious problem, especially in data mining applications that involve the collection and sharing of personal data. For these reasons, the problem of protecting privacy in the context of data mining differs from traditional data privacy protection, as data mining can act as both a friend and foe. Chapter 3 discusses privacy-preserving data mining and its two techniques, namely, those proposed for input data that will be subject to data mining, and those suggested for processed data that are the output of the data mining algorithms. This chapter also presents attacks against the privacy of data mining applications. The chapter concludes with a discussion of next-generation privacy-preserving data mining applications at both the individual and organizational levels.

Chapter 4 explains multi-label classification based on graph neural networks. Typical Laplacian embedding focuses on building Laplacian matrices prior to minimizing weights of connected graph components. However, for multi-label problems, it is difficult to determine such Laplacian graphs owing to multiple relations between vertices. Unlike the typical approaches that require pre-computed Laplacian matrices, this chapter presents a new method for automatically constructing Laplacian graphs during Laplacian embedding. By using trace minimization techniques, the topology of the Laplacian graph can be learned from input data, thus creating robust Laplacian embedding and influencing graph convolutional networks. The experimental results show that the method proposed in this work performs better than the baselines, even when the data is contaminated with noise.

In the cyber world, modern-day malware is quite intelligent with the ability of hiding its presence on the network and performing stealthy operations in the background. Advance persistent threat (APT) is one such kind of malware attack on sensitive corporate and banking networks that can remain undetected for a long time. In real-time corporate networks, identifying the presence of intruders is a challenging task for security experts. Chapter 5 presents a study on data mining and machine learning techniques in APT attribution and detection. In this chapter, the authors shed light on various data mining, machine learning techniques and frameworks used in both attribution and detection of APT malware. Additionally, the chapter highlights gap analysis and the need for a paradigm shift in existing techniques to deal with evolving modern APT malware.

Instagram is one of the world's top ten most popular social networks. One of the main purposes of Instagram is social media marketing. Chapter 6 focuses on text classification of Instagram captions using support vector machine (SVMs). The proposed SVM algorithm uses text classification to categorize Instagram captions into organized groups, namely fashion, food and beverage, technology, health and beauty, lifestyle and travel, and so on, in 66,171 post captions to classify what is trending on the platform. The chapter uses the term frequency-inverse document frequency (TF-IDF) method and percentage variations for data separation in this study.

The main challenges in data mining are related to large, multi-dimensional data sets. There is a need to develop algorithms that are precise and efficient enough to deal with big data problems. The simplex algorithm from linear programming is an example of a successful big data problem-solving tool.

According to the fundamental theorem of linear programming, the solution of the optimization problem can be found in one of the vertices in the parameter space. The basis exchange algorithms also search for the optimal solution among a finite number of vertices in the parameter space. Basis exchange algorithms enable the design of complex layers of classifiers or predictive models based on a small number of multivariate data vectors. Chapter 7 discusses computing on vertices in data mining. The chapter considers computational schemes of designing classifiers or prognostic models based on a data set that consists of a small number of high-dimensional feature vectors. It also discusses in detail the concept of a complex layer composed of many linear prognostic models built in low-dimensional features.

Nowadays, the increase in data acquisition and complexity around optimization make it imperative to jointly use artificial intelligence and optimization for devising data-driven and intelligent decision support systems. A decision support system can be successful if large amounts of interactive data is processed fast to extract useful information and knowledge to help in real-time decision-making. In this context, the data-driven approach has gained prominence due to its provision of insights for decision-making and easy implementation. The data-driven approach can discover various database patterns without relying on prior knowledge while also handling flexible objectives and multiple scenarios. Chapter 8 introduces artificial intelligence and its application in data-driven optimization. The chapter reviews recent advances in data-driven optimization, highlighting the promise of data-driven optimization that integrates mathematical programming and machine learning for decision-making. It also presents perspectives on reinforcement learning (RL)-based data-driven optimization and deep RL for solving NP-hard problems. The chapter investigates the application of data-driven optimization in different case studies to demonstrate the improvements in operational performance over conventional optimization methodology. Finally, the chapter includes some managerial implications and provides some future directions.

Chapter 9 is a detailed discussion on the practical application of the clustering algorithm. This chapter surveys the clustering algorithm, which is an unsupervised learning algorithm for data mining and machine learning techniques. The most popular clustering algorithm is the K-means clustering algorithm, where it is required to find an appropriate K value for distributing the training dataset. It is common to find this value experimentally. Also, it can use the elbow method, which is a heuristic approach used in determining the number of clusters. The particulate matter concentration clustering algorithm for particulate matter distribution estimation performs a K-means clustering algorithm to cluster feature data sets to find the observatory location representing particulate matter distribution.

Chapter 10 looks at the leaching mechanisms of trace elements from coal and host rock using data mining. Coal and host rock, including gangue dump, are important sources of toxic elements that have great potential to contaminate surface and ground water. The leaching and migration of trace elements are controlled mainly by two factors: trace elements' occurrence and surrounding environment. The traditional method to investigate elements' occurrence and leaching mechanisms is based on a geochemical method. In this chapter, data mining is applied to discover the relationship and patterns that are concealed in the data matrix. From the geochemical point of view, the patterns mean the occurrence and leaching mechanisms of trace elements from coal and host rock. An unsupervised machine learning method using principal component analysis is applied to reduce dimensions of the data matrix of solid and liquid samples, then the re-calculated data is clustered to find its co-existing pattern using the Gaussian mixture model.

Chapter 11 introduces the sentiment mining of tourists based on deep learning. Mining the sentiment of the user on the Internet via the context plays a significant role in uncovering human emotion and determining the exactness of the underlying emotion in the context. An increasing number of user-generated content in social media and online travel platforms lead to the development of data-driven sentiment analysis, and most extant SA (sentiment analysis) in the domain of tourism is conducted using document-based SA. However, DBSA (document-based sentiment analysis) cannot be used to examine what specific aspects need to be improved or disclose the unknown dimensions that affect the overall sentiment like aspect-based SA. ABSA (aspect-based sentiment analysis) requires accurate identification of the aspects and sentiment orientation in the UGC (User-generated content). In this chapter, the contribution of data mining based on deep learning in sentiment and emotion detection are clearly illustrated.

Chapter 12 explains data mining applied for predicting community satisfaction of rehabilitation and reconstruction projects. Natural disasters can occur anytime and anywhere, especially in areas with high disaster risk. Rehabilitation and reconstruction projects have been implemented to restore and accelerate economic growth in such cases. As such, a study is needed to determine whether the rehabilitation and reconstruction that has been carried out resulted in community satisfaction. The results of further analysis are expected to predict the level of community satisfaction for the implementation of rehabilitation and other reconstruction. This chapter uses predictive modeling with a data mining approach. The analysis results show that the artificial neural network and the SVM with a data mining approach can develop a community satisfaction prediction model to implement rehabilitation and reconstruction after earthquake-tsunami and liquefaction disasters.

This book is for students, researchers, practitioners, data analysts, and business professionals who seek information on the various data mining techniques and their applications.

I would like to convey my gratitude to everyone who contributed to this book including the authors of the accepted chapters. My special thanks to Publishing Process Manager, Ms. Mia Vulovic, and other staff at IntechOpen for their support and efforts in bringing this book to fruitful completion.

Ciza Thomas, PhD
Professor,
Senior Joint Director,
Directorate of Technical Education,
Government of Kerala,
India

Section 1

Concepts of Data Mining

Chapter 1

The Concept of Data Mining

Julius Olufemi Ogunleye

Abstract

Data mining is a technique for identifying patterns in large amounts of data and information. Databases, data centers, the internet, and other data storage formats; or data that is dynamically streaming into the network are examples of data sources. This paper provides an overview of the data mining process, as well as its benefits and drawbacks, as well as data mining methodologies and tasks. This study also discusses data mining techniques in terms of their features, benefits, drawbacks, and application areas.

Keywords: Data mining techniques, Data mining process, Regression Analysis, Statistical techniques, Clustering techniques, Neural networks, Nearest Neighbors, Decision trees, Rule induction

1. Introduction

1.1 Data Mining

Data mining may be thought of as a natural progression of information technology. It can be simply defined as a procedure for searching, gathering, filtering, and analyzing data. It's the method of extracting useful knowledge from vast volumes of data kept in databases, data centers, or other data repositories. Database technology, statistics, artificial intelligence, machine learning, high-performance computing, pattern recognition, neural networks, data visualization, information retrieval, image and signal processing, and spatial data analysis are all techniques used in data mining. Data mining allows for the extraction of interesting knowledge, regularities, or high-level information from databases, which can then be viewed or browsed from various perspectives.

It deals with the secondary study of massive databases in order to uncover previously unknown relationships that are of interest or benefit to database owners. It can be thought of as computer-assisted exploratory data analysis of massive, complex data sets from a statistical standpoint. Data mining is having a big effect in business, industry, and science right now. It also opens up a lot of possibilities for new methodological advances in science. New issues occur, partly as a result of the sheer scale of the data sets in question, and partly as a result of pattern matching issues.

Decision making, process control, information management, and query processing are only a few of the applications for the newly discovered experience. As a result, data mining is recognized as one of the most exciting modern database technologies in the information industry, as well as one of the most important frontiers in database systems [1, 2]. This chapter will go into the basics of data mining as well as the data extraction techniques. Mastering this technology and its techniques will have significant advantages as well as a competitive edge.

1.2 The importance of data mining

Data mining has gotten a lot of attention in the information industry in recent years because of the widespread availability of massive quantities of data and the pressing need to transform the data into valuable information and knowledge. Business management, quality control, and market research, as well as engineering design and science discovery, will all benefit from the information and expertise acquired. Governments, private corporations, large organizations, and all industries are interested in collecting a large amount of data for business and research purposes [3, 4]. The following are some of the reasons why data mining is so important:

- Data mining is the process of collecting vast amounts of data in order to extract information and dreams from it. The data industry is currently experiencing rapid growth, which has resulted in increased demand for data analysts and scientists.

- We interpret the data and then translate it into useful information using this technique. This enables an organization to make more accurate and better decisions. Data mining aids in the creation of wise business decisions, the execution of accurate campaigns, the prediction of outcomes, and many other tasks.

- We can evaluate consumer habits and insights with the aid of data mining. This results in a lot of growth and a data-driven business.

It's important to remember that which data mining approach to utilize is mostly determined by the amount of data accessible, the type of data, and the dimensions. Although there are evident differences in the types of challenges that each data mining technique is best suited for, the nature of data from the real world, as well as the complicated ways in which markets, customers, and the data that represents them, means that the data is always altering. As a result, there is no obvious law that favors one technique over another. Decisions are sometimes made depending on the availability of experienced data mining analysts in one or more techniques. The preference of one technique over the others depends more on getting good resources and good analysts (**Figure 1**) [5].

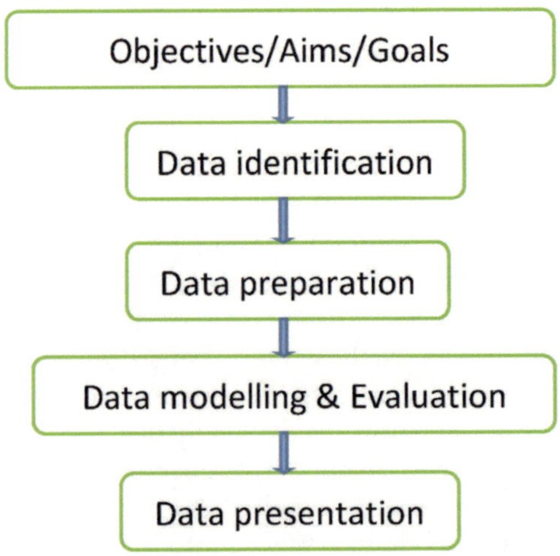

Figure 1.
An overview of data mining process.

2. Related works

2.1 An overview of efficient data mining techniques

Data mining, according to Sandeep Dhawan, is the act of uncovering relationships within large data sets, as well as data trends, anomalies, changes, and significant statistical structures. Forming a hypothesis and then testing it against the dataset are two common data analysis strategies. Data mining techniques, on the other hand, find significant patterns in data automatically, and these patterns can be utilized to construct algorithms. The result or pattern detected should be genuine, intelligible, and valuable, which is a crucial issue when mining large data sets. It goes without saying that the efficiency of robust and intelligent data mining algorithms is essential for data warehousing and sustaining massive datasets. Data mining techniques are being used in practically every sector of the business world. There is rarely any sphere of life that does not have an input and integration of these data mining tools, from the music industry to film maintenance and medicine to sports. The study gave an overview of some of the most widely used data mining techniques, as well as their applications [3].

2.2 An overview of data mining techniques

The study offered an overview of some of the most widely used data mining algorithms. They were divided into two portions, each with its own theme:

- Statistics, Neighborhoods, and Clustering are examples of traditional techniques.

- Trees, Networks, and Rules: Next-Generation Techniques.

The authors discussed a variety of data mining methods so that the reader may see how each algorithm fits into the larger picture of data mining approaches. There were six different types of data mining algorithms presented in all. The Authors noted that, although there are a number of other algorithms and many variations of the techniques that were described, one of the algorithms is almost always used in real-world deployments of data mining systems [4].

3. Methods

Data mining is a multi-stage process that is accomplished in stages. Data mining is now widely employed in a variety of fields. Data mining has applications and uses in almost every aspect of life, and there are a variety of commercial data mining solutions available today [6–8].

3.1 Data mining process

Data mining is a collaborative effort that includes the following steps:

1. Collecting requirements

   The collection and understanding of requirements is the first step in any data mining project. With the vendor's business viewpoint, data mining analysts or users determine the requirement scope.

2. Data investigation

   This move entails identifying and converting data patterns using data mining statistics. It necessitates collecting, assessing, and investigating the requirement or project. Experts comprehend the issues and challenges and translate them into metadata.

3. Data collection and planning

   For the modeling phase, data mining experts translate the data into meaningful information. They use the ETL (Extract, Transform, and Load) method. They're also in charge of inventing new data attributes. Various methods are employed here to view data in a structural format while preserving the value of data sets.

4. Modeling

   Data experts use their best tools for this phase because it is so important in the overall data processing. To filter the data in an acceptable way, all modeling methods are used. Modeling and assessment are intertwined steps that must be completed at the same time to ensure that the criteria are correct. After the final modeling is completed, the accuracy of the final result can be checked.

5. Assessment or Evaluation

   After efficient modeling, this is the filtering method. If the result is not acceptable, it is then passed back to the model. After a satisfactory result, the requirement is double-checked with the provider to ensure that no details are overlooked. At the end of the process, data mining experts evaluate the entire outcome.

6. Deployment

   This is the final stage in the entire process. Data is presented to vendors in the form of spreadsheets or graphs by experts.

The following functions can be performed with data mining services [9, 10]:

- ***Knowledge extraction:*** This is the procedure for finding useful trends in data that can be used in decision-making [11]. This is because decisions must be made on the basis of correct/accurate data and evidence.

- ***Data collection:*** By scraping through linked websites and databases, it is possible to collect information about investors, portfolios, and funds using the web scraping process.

- ***Web data:*** Web data is notoriously difficult to mine. This is due to the essence of the situation. Web data, for example, can be considered dynamic, meaning it changes over time. As a result, the data mining process should be replicated at regular intervals.

- ***Data pre-processing:*** Typically, the data gathered is stored in a data center. This information must be pre-processed. Data mining experts should manually delete any data that is considered unimportant during pre-processing.

- ***Market research, surveys, and analysis:*** Data mining can be used for product research, surveys, and market research. It is possible to collect data that would be useful in the creation of new marketing strategies and promotions.

- ***Scanning of data:*** Data obtained and processed would be useless until it is scanned. Scanning is essential for detecting trends and similarities in the data.

- ***Customer feedback:*** A company's operations are heavily influenced by customer feedback and suggestions. Customers can easily find the details on forums, journals, and other sites where they can openly express their opinions.

- ***News:*** With nearly all major newspapers and news outlets sharing their news online these days, it is easy to collect information on developments and other important topics. It is possible to be in a better place to compete in the market this way.

- **Up-to-date data:** Keeping data up to date is important. The information gathered would be useless unless it is modified. This is to ensure that the data is valid before making decisions based on it.

- ***Internet research:*** The internet is well-known for its vast amount of knowledge. It is obvious that it is the most important source of data. It is possible to collect a great deal of knowledge about various businesses, consumers, and company clients. Frauds can be detected using online resources.

- ***Study of competitors:*** It's important to know how your competitors are doing in the business world. It is important to understand both their strengths and weaknesses. Their methods of marketing and distribution can be mined, including their methods of reducing overall costs.

3.2 Advantages of data mining

Data mining and its features have many advantages. It raises the need for a data-driven market as it is combined with analytics and big data. Some of the benefits are as follows:

1. Manufacturing industries benefit from data mining by detecting defective devices and goods using engineering data. This aids in the removal of defective goods from the stock list.

2. It assists government agencies in analyzing financial data and transactions in order to model them into usable data.

3. Data mining is useful not only for making forecasts, but also for developing new services and goods.

4. Predictive models are used in the retail sector for products and services. Better quality and consumer insights are possible in retail stores. Historical data is used to calculate discounts and redemption.

5. Data mining aids financial gains and alerts for banks. They create a model based on consumer data that aids in the loan application process, among other things.

6. Customers gain confidence in companies, which leads to an increase in the number of clients.

7. Marketing firms use data mining to create data models and forecasts based on historical data. They manage promotions, marketing strategies, and so on. This leads to fast growth and prosperity.

8. Data mining results in the creation of new revenue sources, resulting in the expansion of the company.

9. Data mining aids in the improvement of strategy and decision-making processes in organizations.

10. When competitive advantages are found, data mining can help reduce production costs.

3.3 Data mining techniques and tasks

Understanding the types of tasks, or problems, that data mining can solve is the best way to learn about it. The majority of data mining tasks can be classified as either prediction or summary at a high level. Predictive tasks allow you to forecast the value of a variable based on previously collected data. Predicting when a customer will leave a business, predicting whether a transaction is fraudulent, and recognizing the best customers to receive direct marketing offers are all examples of predictive tasks. Descriptive tasks, on the other hand, attempt to summarize the information. Automatically segmenting customers based on their similarities and differences, as well as identifying correlations between products in market-basket data, are examples of such tasks [12].

Organizations now have more data at their disposal than they have ever had before. However, due to the sheer volume of data, making sense of the massive amounts of organized and unstructured data to enact organization-wide changes can be exceedingly difficult. This problem, if not properly handled, has the potential to reduce the value of all the data.

Data mining is the method by which businesses look for trends in data to gain insights that are important to their needs. Both business intelligence and data science need it. Organizations may use a variety of data mining strategies to transform raw data into actionable insights [13]. These range from cutting-edge artificial intelligence to the fundamentals of data planning, all of which are critical for getting the most out of data investments.

a. Cleansing and preparing data

b. Pattern Recognition

c. Classification

d. Association

e. Detection of Outliers

f. Clustering

g. Regression

h. Prediction

i. Sequential trends

j. Decision Trees

k. Statistical techniques

l. Visualization

m. Neural Networks

n. Data warehousing

o. Machine Learning and Artificial intelligence

a. Cleansing and preparing data

Cleaning and preparing data is a vital part of the data mining process. Raw data must be cleansed and formatted in order to be useful in various analytic approaches. Different elements of data modeling, transformation, data migration, ETL, ELT, data integration, and aggregation are used in data cleaning and planning. It's a necessary step in determining the best use of data by understanding its basic features and attributes. Cleaning and preparing data has obvious business value. Data is either useless to an entity or inaccurate due to its accuracy if this first phase is not completed. Companies must be able to trust their data, analytics results, and the actions taken as a result of those findings. These measures are also needed for good data quality and data governance.

b. Pattern Recognition

A basic data mining technique is pattern recognition. It entails spotting and tracking trends or patterns in data in order to draw informed conclusions about business outcomes. When a company notices a pattern in sales data, for example, it has a reason to act. If it's determined that a certain product sells better than others for a specific demographic, a company may use this information to develop similar goods or services, or simply better stock the original product for this demographic [14].

c. Classification

The various attributes associated with different types of data are analyzed using classification data mining techniques. Organizations may categorize or classify similar data after identifying the key characteristics of these data types. This is essential for recognizing personally identifiable information that organizations may wish to shield or redact from records, for example.

d. Association

The statistical technique of association is a data mining technique. It denotes that some data (or data-driven events) are linked to other data. It's similar to the machine learning concept of co-occurrence, where the existence of one data-driven event indicates the probability of another. Correlation and association are two statistical concepts that are very similar. This means that data analysis reveals a connection between two data occurrences, such as the fact that hamburger purchases are often followed by French fries purchases.

e. Detecting of Outliers

Outlier detection is used to identify the deviations in datasets. When companies discover anomalies in their records, it becomes easier to understand why they occur and plan for potential events in order to achieve business goals. For example, if there is an increase in the use of transactional systems for credit cards at a certain time of day, businesses can use this information to maximize their income for the day by finding out the cause of it.

f. Clustering

Clustering is an analytics methodology that employs visual approaches to data interpretation. Graphics are used by clustering mechanisms to demonstrate where data distribution is in relation to various metrics. Different colors are used in clustering techniques to represent data distribution. When it comes to cluster analytics, graph-based methods are perfect. Users can visually see how data is distributed and recognize patterns related to their business goals using graphs and clustering in particular.

g. Regression

The essence of the relationship between variables in a dataset can be determined using regression techniques. In some cases, such connections may be causal, and in others, they may only be correlations. Regression is a simple white box technique for revealing the relationships between variables. In areas of forecasting and data modeling, regression methods are used (**Figure 2**).

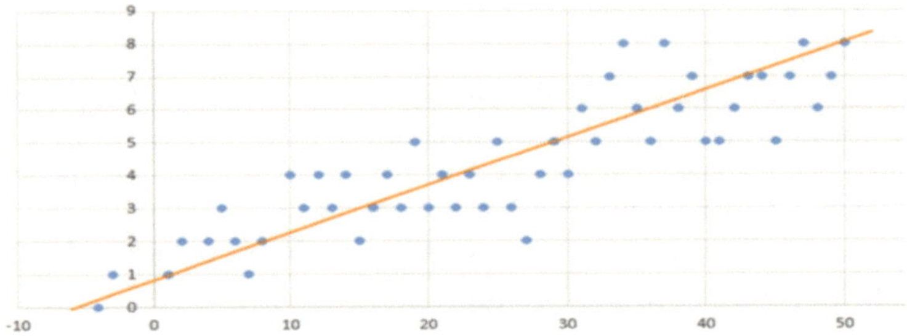

Figure 2.
Illustration example of linear regression on a set of data [15].

h. Prediction

One of the four branches of analytics is prediction, which is a very important feature of data mining. Patterns observed in current or historical data are extended into the future using predictive analytics. As a result, it allows businesses to predict what data patterns will emerge next. Using predictive analytics can take a variety of forms. Machine learning and artificial intelligence are used in some of the more advanced examples. Predictive analytics, on the other hand, does not have to rely on these methods; simpler algorithms can also be used.

i. Sequential Trends

This data mining technique focuses on identifying a sequence of events. It's particularly useful for transactional data mining. For example, when a customer buys a pair of shoes, this technique will show which pieces of clothing they are more likely to buy. Understanding sequential trends may assist businesses in recommending additional products to consumers in order to increase sales.

j. Decision trees

Decision trees are a form of predictive model that enables businesses to mine data more effectively. A decision tree is technically a machine learning technique, but because of its simplicity, it is more often referred to as a white box machine learning technique. Users can see how the data inputs influence the outputs using a decision tree. A random forest is a predictive analytics model that is created by combining different decision tree models. Complicated random forest models are referred to as "black box" machine learning techniques because their outputs are not always easy to comprehend based on their inputs. However, in most cases, this simple form of ensemble modeling is more effective than relying solely on decision trees (**Figure 3**).

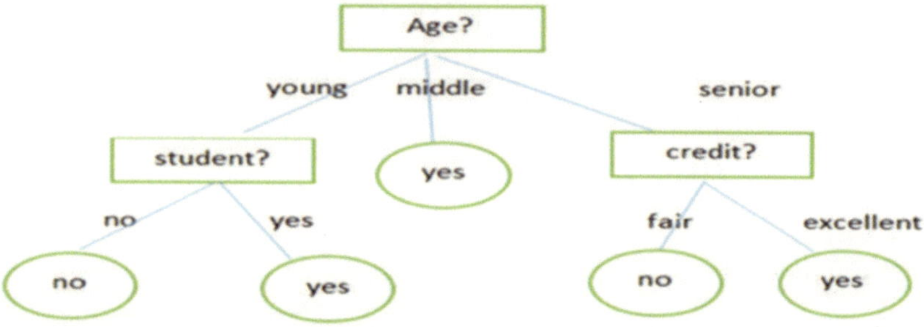

Figure 3.
Example of a decision tree [15].

k. Statistical techniques

Statistical approaches are at the heart of the majority of data mining analytics. The various analytics models are focused on mathematical principles that produce numerical values that can be used to achieve clear business goals. In image recognition systems, neural networks, for example, use complex statistics based on various weights and measures to decide if a picture is a dog or a cat. Statistical models are one of artificial intelligence's two primary branches. Some mathematical methods have static models, while others that use machine learning improve over time.

l. Visualization

Another essential aspect of data mining is data visualization which uses sensory impressions that can be seen to provide users with access to data. Today's data visualizations are interactive, useful for streaming data in real-time, and distinguished by a variety of colors that show various data trends and patterns. Dashboards are a valuable tool for uncovering data mining insights using data visualizations. Instead of relying solely on the numerical results of mathematical models, organizations may create dashboards based on a variety of metrics and use visualizations to visually illustrate trends in data.

m. Neural Networks

A neural network is a type of machine learning model that is frequently used in AI and deep learning applications. Among the most accurate machine learning models used today is neural network. They are named for the fact that they have multiple layers that resemble how neurons function in the human brain. While a neural network can be a powerful tool in data mining, companies should exercise caution when using it because some of these neural network models are extremely complex, making it difficult to understand how a neural network calculated an output (**Figure 4**).

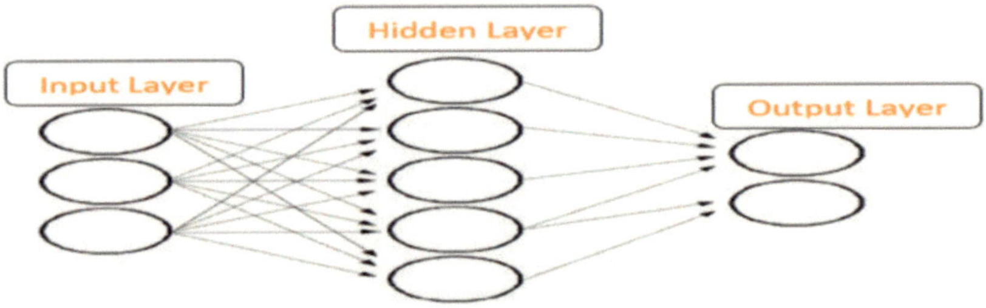

Figure 4.
Example of a neural network [15].

n. Data warehousing

Data warehousing used to imply storing organized data in relational database management systems so that it could be analyzed for business intelligence, reporting, and simple dashboarding. Cloud data centers and data warehouses in semi-structured and unstructured data stores, such as Hadoop, are available today. Although data warehouses have historically been used to store and analyze historical data, many new approaches can now provide in-depth, real-time data analysis.

o. Machine Learning and Artificial Intelligence

Some of the most advanced advances in data mining are machine learning and artificial intelligence (AI). When operating with large amounts of data, advanced machine learning techniques such as deep learning provide extremely accurate predictions. As a result, they can be used to process data in AI applications such as computer vision, speech recognition, and advanced text analytics using Natural Language Processing. These data mining techniques work well with semi-structured and unstructured data to determine meaning.

3.4 Data mining software for optimization

With so many methods to use during data mining, it's important to have the right resources to get the most out of your analytics. For proper implementation, these methods usually necessitate the use of many different tools or a tool with a broad set of capabilities.

While organizations can use data science tools like R, Python, or Knime for machine learning analytics, it's critical to use a data governance tool to ensure compliance and proper data lineage. Additionally, in order to conduct analytics, companies would need to collaborate with repositories such as cloud data stores, as well as dashboards and data visualizations to provide business users with the knowledge they need to comprehend analytics. All of these features are available in tools, but it's critical to find one or more that meet your company's requirements [16].

4. Discussions

4.1 The cloud and data mining's future

The development of data mining has been accelerated by cloud computing technology. Cloud systems are ideally adapted for today's high-speed, massive amounts of semi-structured and unstructured data that most businesses must contend with. The elastic capabilities of the cloud will easily scale to meet these big data demands. As a result, since the cloud can carry more data in a variety of formats, more data mining techniques are needed to transform the data into insight. Advanced data mining techniques such as AI and deep learning are now available as cloud services.

Future advancements in cloud computing would undoubtedly increase the need for more powerful data mining software. AI and machine learning will become even more commonplace in the next five years than they are now. The cloud is the most suitable way to both store and process data for business value, given the exponentially growing pace of data growth on a daily basis. As a result, data mining methods can depend much more on the cloud than they do now.

Currently, data scientists use a variety of data mining techniques, which differ in precision, efficiency, and the type and/or volume of data available for analysis. Classical and modern data mining techniques are two types of data mining techniques. Statistical approaches, Nearest Neighbors, Clustering, and Regression Analysis are examples of Classical techniques, while Modern techniques include Neural Networks, Rule Induction Systems, and Decision Trees.

4.2 Data mining techniques – advantages and disadvantages

1. Statistical Techniques

Advantages

- Because secondary data is normally inexpensive and requires less time to compile because it has already been done.
- Because the patterns and correlations are obvious and reliable.
- Broad samples were used to ensure high generalizability.
- It can be used several times to test various variables.
- It is possible to assess changes that enhance reliability and representativeness.

Disadvantages

- The researcher is only able to draw patterns and correlations from the data and cannot assess the validity or consider a causal theory process.
- Statistical data is often secondary data, making misinterpretation simple.
- Statistical evidence can be manipulated, and it can be skewed and phrased to support the researcher's point (effects objectivity).
- It's difficult to view and validate this data because it's always secondary.

2. Nearest Neighbors

Advantages

- It's easy and intuitive to use
- It does not make any assumptions
- No Education Transfer
- It is constantly growing.
- It is a simple multi-class issue to enforce.
- Regression and classification are also possible applications.
- Selecting the first hyper parameter can take some time, but once done, the rest of the parameters are compatible with it.
- Provides a variety of distance parameters to choose from (Euclidean distance, Hamming distance, Manhattan distance, and Minkowski distance).

Disadvantages

- Irrelevant characteristics can influence the distance between neighbors.
- While the implementation can be simple, the efficiency (or speed of the algorithm) decreases rapidly as the dataset grows.
- Can handle a small number of input variables, but as the number of variables increases, the algorithm has trouble predicting the performance of new data points.
- Characteristics must be consistent.
- When classifying new data, the problem of determining the optimal number of neighbors to consider is frequently encountered.
- Issues with using data that is unbalanced.
- It is vulnerable to outliers since neighbors are clearly selected based on distance parameters.

3. Clustering

Advantages

- Hierarchical methods enable the end user to choose from a large number of clusters or a small number of clusters.
- Appropriate for data sets of any form and attributes of any kind.
- There are a variety of well-developed models that provide a way to accurately represent the data, and each model has its own unique characteristics that can provide significant advantages in some specific areas.

Disadvantages

- Cluster numbers must be preset.
- The assumption is not completely right, and the clustering result is dependent on the parameters of the chosen models.

4. Regression Analysis *(MARS- Multivariate Analysis for Regression Splines, OLS - Ordinary Least Square regression, SVR-Support Vector Regression, Radial Basis Function Networks)*

Advantages

- Linear regression can solve some very simple problems much faster and more easily, since prediction is simply a multiple of the predictors.
- Linear regression: the modeling process is simple, requires few calculations, and runs quickly even when the data is large.
- Linear regression: the factor can provide insight into and interpretation of each variable.
- In linearly separable datasets, linear regression works well.
- Linear regression is easier to implement, evaluate, and apply than other methods.
- In linear regression, dimensionality reduction, regularization (L1 and L2), and cross-validation methods can all be used to prevent over-fitting.
- Multiple regression will assess the relative importance of one or more predictor variables in determining the criterion's significance.
- Outliers, or deviations, can be found using multiple regression.

Disadvantages

- Linear regression: There is a minimum linear association.
- Linear regression: Outliers are affected easily.
- The regression solution would most likely be thick (because there is no regularization)
- Linear regression is vulnerable to noise and overfitting.
- Regression solutions obtained through a variety of approaches (e.g., optimization, less-square, QR decomposition, etc.) are not necessarily unique.
- Vulnerable to multicollinearity: Multicollinearity should be eliminated (using dimensionality reduction techniques) before using linear regression since it means that the independent variables have no relationship.

- Any disadvantage of using a multiple regression model is usually due to the data used, either because there is insufficient data or because the cause is incorrectly assumed to be a correlation.

5. Neural Networks

Advantages

- Artificial Neural Networks (ANN) will model and analyze nonlinear, complex relationships.
- Has highly accurate statistical models that can be used to solve a wide range of problems.
- Information is stored on the network as a whole, not in a database, and the network will run even though a few pieces of information are missing from one location.
- The ability to work with limited knowledge
- Has fault tolerance, which means that contamination of one or more ANN cells will not stop development.
- Is endowed with a memory.
- Gradual corruption: As a network ages, it slows down and becomes more vulnerable. The network issue does not seem to be corroding right away.
- Machine-training capability: Artificial neural networks learn events by observing and reflecting on similar events.
- Parallel processing capability: Artificial neural networks have the computing capacity to perform several tasks at the same time.

Disadvantages

- Extraction of features-the issue of determining which predictors are the most suitable and significant in building models that are predictably accurate.
- Hardware reliance: Artificial neural networks, by their very nature, require parallel processing processors. This is the foundation on which the equipment realization is built.
- Assurance of proper network structure: When it comes to artificial neural network design, there are no hard and fast rules. The correct network design is achieved by practice and trial and error.
- Network behavior that is not explained: Even though ANN provides a sampling solution, it does not explain why or how it works.
- Difficulty in demonstrating the problem to the network: ANNs should deal with numerical data. Before integrating into ANN, problems must be translated into numerical values.

- The network's length is unknown: reducing the network to a certain sample error value indicates that the training is complete. This may not result in optimal results.

6. Rule Induction

Advantages

- When dealing with a small number of rules, IF-THEN rules are easy to understand and are meant to be the most interpretable model.
- The decision rules are just as descriptive as decision trees, but they are a lot smaller.
- Since only certain conditional statements must be checked to determine the rules apply, IF-THEN rules are simple to predict.
- Since conditions only shift at the threshold, decision rules will withstand monotonous input function transformations.
- IF-THEN rules produce models with few features. Only the features that are important to the model are chosen.
- Simple rules like OneR can be used to test more complex algorithms.

Disadvantages

- IF-THEN laws are mostly concerned with grouping and almost completely neglect regression.
- Categorical functions are also needed. This means that numerical features must be classified if they are to be included.
- The majority of older rule-learning algorithms are prone to overfitting.
- In the study of linear feature-output relations, decision rules are ineffective.

7. Decision Trees *(CART – Classification and Regression Trees)*

Advantages

- Data is organized into distinct categories, which are therefore simpler to grasp than points on a multidimensional hyperplane, as in linear regression. With its nodes and edges, the tree structure has a natural visualization.
- In real-world problems, the models to be built and the interactions to be detected are usually much more complex.
- CART validates the Tree immediately, implying that the algorithm has the model validation and discovery of the optimally general model (the algorithm) built deep inside it.
- When it comes to missing data, the CART algorithm is fairly reliable.
- There are so many powerful data mining features that decision trees mark so strongly.

Disadvantages

- Can struggle with some very simple problems where prediction is simply a multiple of predictors.
- Trees are incapable of handling linear relationships. Splits must be used to approximate any linear input–output relationship, resulting in a phase function. This is not going to work.
- It had a silky feel to it. Small changes in the input function may have a big effect on the forecast outcome, which is not always a good thing.
- The trees are still very shaky. A few tweaks to the training dataset will result in a completely different tree. Since every split is based on splitting the parent, this is the case.

These methods are best applied to particular tasks in order to achieve the best performance. The **Table 1** below lists the data mining tasks and the techniques that can be used to complete them.

A business analyst's dream is data warehousing. All of the data concerning the organization's actions is centralized and accessible through a single set of analytical tools. A data warehouse system's goal is to give decision-makers the accurate, timely data they need to make the best decisions possible. A relational database management system server serves as the central repository for informational data in the data warehouse architecture. The processing of operational data is kept distinct from the processing of data warehouse data.

The central information repository is surrounded by a number of critical components that work together to make the overall ecosystem functional, manageable, and available to both operational systems and end-user query and analysis tools. The warehouse's raw data is often derived from operational applications. Data is cleansed and turned into an integrated structure and format when it enters the warehouse. Conversion, summarization, filtering, and condensing of data may all be part of the transformation process. Since the data contains a historical component, the warehouse must be capable of holding and managing large volumes of data as well as different data structures for the same database over time.

No	Data mining task	Data mining techniques
1	Classification	Decision trees, Neural networks, K-nearest neighbors, Rule induction methods, SVM-Support vector machine, CBR-Case based reasoning
2	Prediction	Neural networks, K-nearest neighbors, Regression Analysis
3	Dependency Analysis	Correlation analysis, Regression Analysis, Association rules, Bayesian networks, Rule Induction
4	Data description and summarization	Statistical techniques, OLAP (Online Analytical Processing)
5	Segmentation or clustering	Clustering techniques, Neural Networks
6	Consolidation	Nearest neighbors, Clustering

Table 1.
Data mining tasks and the methods used to accomplish them.

The following **Table 2** lists data mining techniques and their areas of applications.

Data mining techniques	Areas of use
Association Analysis	Designing store shelves, marketing, cross-selling of products.
Classification (K-nearest neighbor, etc.)	Banks, marketing campaign designs by organizations.
Decision Trees	Medicine, engineering, manufacturing, and astronomy, to name a few fields. They were used to solve problems ranging from credit card depletion estimation to time series exchange rate estimation for a variety of international currencies.
Clustering Analysis	Image recognition, web search, and security.
Outlier Detection	Detection of credit card fraud risks, novelty detection, etc.
Regression Analysis (K-nearest neighbor, ...)	Marketing and Product Development Efforts comparison.
Artificial Neural networks	Data compression, feature extraction, clustering, prototype formation, function approximation or regression analysis (including prediction time series, fitness approximation, and modeling), classification (including pattern and sequence recognition, novelty detection, and sequential decision making), data processing (including filtering, clustering, blind source separation, and compression), and robotic compression.
Support vector machines regression	Oil and gas industry, classification of images and text and hypertext categorization.
Multivariate Regression algorithm	Retail sector
Linear Regression	Financial portfolio prediction, salary forecasting, real estate predictions and in traffic estimated time of arrivals (ETAs).

Table 2.
Data mining techniques and their areas use.

5. Conclusion

It's worthy of note to state that time is spent on extracting useful information from data. As a result, in order for companies to develop quickly, it is necessary to make accurate and timely decisions that enable them to take advantage of available opportunities. In today's world of technology trends, data mining is a rapidly growing industry. In order to obtain valuable and reliable information, everyone today needs data to be used in the right way and with the right approach. Data mining can be initiated by gaining access to the appropriate resources. Since data mining begins immediately after data ingestion, finding data preparation tools that support the various data structures required for data mining analytics is important. Organizations may also want to identify data in order to use the aforementioned methods to explore it. Modern data warehousing, as well as various predictive and machine learning/AI techniques, are helpful in this regard.

Choosing which approach to employ, and when, is clearly one of the most difficult aspects of implementing a data mining process. Some of the parameters that are critical in deciding the technique to be used are determined by trial and error. There are clear differences in the types of problems that each data mining technique is best suited for. As a result, there is no simple rule that favors one technique over another. Decisions are often taken based on the availability of qualified data mining analysts in one or more techniques. The choice of a technique over the other is more dependent on the availability of good resources and analysts.

Acknowledgements

This work was supported by the Faculty of Applied Informatics, Tomas Bata University in Zlín, Czech Republic, under Projects IGA/CebiaTech/2021/001.

Author details

Julius Olufemi Ogunleye
Tomas Bata University in Zlin, Zlín, Czech Republic

*Address all correspondence to: juliusolufemi@yahoo.com

References

[1] Software Testing Help (April 16, 2020): Data Mining Techniques: Algorithm, Methods & Top Data Mining Tools.

[2] Silhavy, P., Silhavy, R., & Prokopova, Z. (2019): Categorical variable segmentation model for software development effort estimation. IEEE Access, 7, 9618-9626.

[3] Sandeep Dhawan (2014): An Overview of Efficient Data Mining Techniques.

[4] Alex Berson et al. (2005): An Overview of Data Mining Techniques.

[5] Jiawei H. and Micheline K. (2000): Data Mining: Concepts and Techniques.

[6] ACM SIGKDD (2006-04-30), Retrieved (2014-01-27): Data Mining Curriculum.

[7] Kamber H. Et al. (2011): Data Mining: Concepts and Techniques (3rd ed.). Morgan Kaufmann. ISBN 978-0-12-381479-1.

[8] Clifton C. (2010): Encyclopædia Britannica: Definition of Data Mining". Retrieved 2010-12-09.

[9] Weiss G. M. and Davison B. D. (2010): Data Mining (Handbook of Technology Management, H. Bidgoli (Ed.), John Wiley and Sons, 2010).

[10] Mehmed K. (2011): Data Mining Concepts, Models, Methods, and Algorithms (Second Edition).

[11] Berson A. et.al (2005): An Overview of Data Mining Techniques (Excerpts from the book by Alex Berson, Stephen Smith, and Kurt Thearling).

[12] Karna H. et al. (2018): Application of data mining methods for effort estimation of software projects.

[13] Sehra S.K. et al. (2014): Analysis of Data Mining techniques for software effort estimation.

[14] Trevor H. Et al.(2009): The Elements of Statistical Learning: Data Mining, Inference, and Prediction. Archived from the original on 2009-11-10. Retrieved 2012-08-07.

[15] Ogunleye J.O. (2020): Review of Data Mining Techniques in Software Effort Estimation.

[16] Dejaeger K., et al.(2012): Data Mining Techniques for Software Effort Estimation: A Comparative Study.

Chapter 2

Use Data Mining Cleansing to Prepare Data for Strategic Decisions

Mawande Sikibi

Abstract

Pre-processing data on the dataset is often neglected, but it is an important step in the data mining process. Analyzing data that has not been carefully screened for such challenges can produce misleading results. Thus, the representation and quality of data are first and foremost before running an analysis. In this paper, the sources of data collection to remove errors are identified and presented. The data mining cleaning and its methods are discussed. Data preparation has become a ubiquitous function of production organizations – for record-keeping and strategical making in supporting various data analysis tasks critical to the organizational mission. Despite the importance of data collection, data quality remains a pervasive and thorny challenge in almost any production organization. The presence of incorrect or inconsistent data can significantly distort the results of analyses, often negating the potential benefits of strategical making driven approaches. This tool has removed and eliminated errors, duplications, and inconsistent records on the datasets.

Keywords: Data, Data cleaning, Data collection, Data mining, Data preparation, Data collection, Data quality, Messy data

1. Introduction

Time has changed for the production organizations who believe keeping messy data saves their day. This messy data is in the dataset, which is stored in databases, repositories, and data warehouses. Massive amounts of data are available on their resources for the organization to influence their strategic decision. Data collected from various resources are messy, and this affects the quality of the data result. Data preparation offers a better data quality, which will help the organizations yearly, making most existing methods no longer suitable for messy data.

The growing enthusiasm of messy data on the dataset for data-driven strategic decision-making has created the importance of preparing data for future use over the years. The rapid growth of messy data drives new opportunities for the organization and processing the quality of the data by cleaning and preparing data becomes essential for analysts and users. Unfortunately, this could be handled correctly as reliable data could lead to a misguided strategic decision.

Data mining is no longer a new research field [1]. It aims to prepare data to improve data quality before processing by identifying and removing errors and inconsistencies on the dataset [2]. Data mining can pull data to prepare it to inform organization strategic decisions. However, preparing data can be used before for specific organizational purposes.

Data mining could be added to a single application to pull anomalies within a large dataset. Utilizing the software arranges data in the large dataset to develop efficient organizational strategies. Data mining software is a user-friendly interface that allows organizational analysts and users who may not be technically advanced to execute data preparation in data mining [3]. Putting this capability in the hands of the non-technical user allows responding to data quality issues quickly.

Data preparation is the feature within data mining; it has immeasurable value working with data [4]. Utilizing the software will begin to embed within the organization. Data mining software is available on the market for an organization to use their data in the dataset. Thus, markets are different from a decade ago due to rapid change in the world economy and technological advancement. This technology is popular with marketers because it allows analysts and users to make smart strategic decisions. It enables better development of market strategies for competitive advantage ahead amongst organizations. As vendors continue to introduce solutions, the marketing strategy improves the data quality of the dataset stored in their resources. With data mining, analysts and users can access the dataset in preparation for it to be available for future use.

2. Objectives

Understanding the better contribution of data mining makes to the dataset. In addressing this, an attempt of the following should be met.

- To discover errors and inconsistencies in the dataset for data preparation.
- Minimizes the errors and inconsistencies on the dataset.
- Utilize the use of datasets stored in their various resources.

This paper aims to develop a process data mining capability undertake on the dataset. The literature review considers current knowledge contributions to this topic towards these paper objectives.

3. Literature review

Data preparation corrects inconsistent data in the dataset to prepare quality data [5]. Research indicates that data preparation in data mining formulates a workflow process covering steps to prepare data [6]. However, some research suggested that data preparation begins with data collection to check data quality [7]. This paper aims to demonstrate the evolution of collecting data into preparation steps to influence data quality. The paper examines the data preparation in data mining processes through data collection.

3.1 Data collection

Data mining is often described as an add-on software in checking the data quality in the dataset by searching through the large amount of data stored in databases, repositories, and data warehouses. The data stored is believed to be too messy, inconsistent, and have errors; it is unclear information to analysts and users, make it difficult to be ready to be used for its specific purposes [8]. Overloaded data limit analysts and users; thus, software such as data mining is developed to solve this challenge through automation.

The data mining software uses recognition technologies and statistical techniques to clean messy data and discover the possible rule to govern data in databases, repositories, and data warehouses. Data mining considers the process that requires goals and objectives to be specified [9]. Once the intended goals met, it is necessary to determine what data is collected or available. However, before data is used, data preparation is performed, making data ready for its purposes.

The concept that strategic or effective decisions are based on appropriate data is not new. Finding the correct data for strategic decisions began 30 years ago [10]. During the late 1960s, organizations create reports from production sensors into databases, repositories, and data warehouses. These resources stored data to retrieved and manipulate to produce constructive reports containing information to meet specific strategic decision needs.

In the 1980s, analysts and users began to need data more frequently and to be more individualized. Thus, the organizations started to request data in the resources. Later in the 1990s, analysts and users required immediate access to be more detailed information. This meant to correlate with production and strategic decisions processes. It has helped the analysts and users extract its data from databases, repositories, and data warehouses.

The analysts and users began to realize the need for more tools to prepare data for future uses. Additionally, the organizations recognized the accumulated amount of data; thus, new tools to prepare data before meeting their needs. Such tools enabled the system to search for any possible errors and inconsistencies in the dataset. Data mining software was the first developed to help analysts and users to find quality data from a voluminous amount of data. Because the massive volume of data keeps rapidly growing, preparation methods are urgently needed. Therefore, data mining has become an increasingly important research field [11].

3.2 Data cleansing

Data cleansing is an operation within data mining software that can be performed on the existing data to remove anomalies and obtain the data collection. It involves removing the errors, inconsistencies and transform data into a uniform format in the dataset [12]. With the amount of data collected, manual data cleansing for preparation is impossible as it is time-consuming and prone to errors. The data cleansing process consists of several stages: detecting data errors and repairing the data errors [13]. Although, it is thought of as a tedious exercise. However, establish a process and template for the data cleansing process gives assurance that the method applied is correct. Hence, data cleansing focuses on errors beyond small technical variations and constitutes a significant shift within [14].

Data cleansing based on the knowledge of technical errors expects normal values on the dataset. Missing values may be due to interruptions of the data flow. Hence, predefined rules for dealing with errors and true missing and extreme values are part of better practice. However, it is more efficient to detect the errors by active searching for them on the dataset in a planned way. Lack of data through data cleansing will arise if the analysts and users do not fully understand a dataset, including skips and filters [14].

Moore and McCabe [15] emphasized the serious strategic decision error would endure if the data quality were poor, leading to low data utilization efficiency. Although data cleansing follows data collection, data thoroughly checked for errors, and other inconsistencies are corrected for future use [16]. Although the importance of data-handling procedure is being underlined in better clinical practice and data management guidelines, gaps in knowledge about optimal data handling methodologies and standard of quality data are still present [14].

Detecting and correcting corrupted or inaccurate records help to meet standard quality data from the dataset. Find the incorrect, inaccurate, or irrelevant parts of the data, replace, modify, and delete coarse data [14]. The reality of the matter, data cannot always be used as it is and needs preparation to be used. Achieving higher preparation data quality during a data cleansing process is required to remove anomalies. Thus, the data cleansing process can be defined as assessing data's correctness and improving it. Therefore, enhancing data quality, pre-processing data mining techniques are used to understand the data and make it more easily accessible.

3.3 Data validation

Data validation is described as the process of ensuring data has undergone cleaning to ensure that it is both correct and useful. Although, it intended to provide a guarantee for the fitness and consistency of data in the dataset. Failure or omission in data validation can lead to data corruption. Catching data early on the dataset is important as it helps debug the roots of the cause and roll back in the working state [17]. Moreover, it is important to rely on mechanisms specific to data validation rather than on the detection of second-order effects.

Errors are bound to happen during the data collection process, while data is seldom 100% correct. Data validation helps to minimize erroneous data from the dataset. Data validation rules help organizations follow standards that make it efficient to work with data. Although, duplication data provide challenges to many organizations. Factors that cause the duplication of data are the data entry of machines and operators from production to capture data. An organization needs a powerful matching solution to overcome this challenge of duplicating records to ensure clean and usable data.

Data validation checks the accuracy and data quality of source data, usually performed before processing the data. It can be seen as a form of data cleansing. Data validation ensures that the data is complete (no blanks or empty values), unique (includes different values that are not repeated), and the values that range consistent with the expectations. When moving and merging data, it is important to ensure that data from different sources and repositories conform to organizational rules and not become corrupted due to inconsistencies in type or context. Data validation is a general term and can be performed on any data. However, including data within a single application, such as Microsoft Excel, or merging simple data within a single data store.

The data validation process is a significant aspect of filtering the large dataset and improving the overall process's efficiency. However, every technique or process consists of benefits and challenges; therefore, it is crucial to have a complete acknowledgement. Data handling can be easier if analysts and users adapt this technique with the appropriate process, then data validation can provide the best outcome possible for data. Data validation can be broken down into the following categories: data completeness and data consistency.

3.3.1 Data integrity

Data integrity refers to the integrity of the data. However, for the data to be valid, there should not be any gaps or missing information for data to be truly complete. Occasionally incomplete data is unusable, but it is usually used in the absence of information, leading to cost error and miscalculations.

An incomplete data is usually the result of unsuccessful data collection. This denotes the degree to which all required data are available in the dataset [18]. A measure of data completeness would be the percentage of missing data entries. However, the true goal of data completeness is not to have perfect 100% data. It ensures that data the essential to the purpose of validity. Therefore, it is a necessary component of the data quality framework and is closely related to validity and accuracy.

3.3.2 Data consistency

Data consistency means that there is consistency in the measurement of variables throughout the datasets. This becomes a concern, primarily when data aggregates from multiple sources. Discrepancies in data meanings between data sources can create inaccurate, unreliable datasets. Since the data inconsistency comes from the storage format, semantic expressions, and numerical values, a method of consistent quantification assesses the degree of data consistency quantitatively after defining the degree of consistency.

Data consistency could be the difference between great business success or failure. Data is the foundation for successful organizational strategic decisions, and inconsistent data can lead to misinformed business decisions. Organizations must ensure data consistency, especially when aggregating data from multiple internal or external sources without changing their structure, to be confident and successful in their strategic decision-making.

Data consistency checks that the data values of all instances of the application are the same. These data belong together and describe a specific process at a specific time, which means that the data remains unchanged during processing or transmission. Synchronization and protection measures help to ensure that data consistency during the multi-stage processing [19]. Data consistency is essential to the operation of programs, systems, applications, and databases. Locking measures prevent data from being altered by two applications simultaneously and ensure correct processing order. Controlling simultaneous operations and handling incomplete data are essential to maintain and restore data consistency in power failures.

3.4 Data preparation

Data preparation is the process of cleaning and transforming raw data before processing and analysis for future use. It is an important step before processing and often involves reformatting data, correcting data, and combining data sets to enrich data [20]. Its task is to blend, shape, clean, consolidate data into one file or data table to get it ready for analytics or other organizational purposes.

The data must be clean, formatted, and transformed into something digestible by data mining software to achieve the final preparation stage. These actual processes include a wide range of steps, such as consolidating or separating fields and columns, changing formats, deleting unnecessary or junk data, and making corrections to data.

In this literature review, several studies have used data preparation and data mining on the messy data on the dataset for future use, few studies on the quality data check. This is the gap in this paper, as it aims at reviewing the available data mining preparing methods for messy data. Since the data preparation framework needs to meet data quality criteria, using a quality dimension includes accuracy, completeness, timeliness, and consistency [21]. Quality data check is crucial because it automates data and provides information about the number of valid, missing, and mismatched values in each column. The result shows the quality data above each column in the dataset. A data mining software will help remove errors and inconsistencies in the dataset to meet quality data check percentage [22].

Quality data check on the dataset, it may be better to use a transformation. These quality data checks can create data quality rules which persist in checking columnar data against defined. Performing variety checks, transform data automatically show the effect of transformations on the overall quality of data. It can provide various services for the organization and only with high-quality data and achieve the top-service in the organization [13].

4. Methodology

This chapter aims to provide the research methodology roadmap designed to meet the objectives of this paper. It is important to select an appropriate method to ensure the accuracy, validity, and quality of data and findings. This chapter shows the method chosen, the tools used to extract data and data analysis. Hence, the phenomenological concept is focused on preparing data and reference [23]. A research method refers to how data can be collected and analyzed, such as data analysis software.

This paper used ethnography as the researcher was directly involved in preparing messy data on the dataset. Ethnography is usually described as participant observation, and this was where the researcher became actively involved, demonstrating the data preparation.

A single case approach was chosen for this paper to be the suitable method for executing data preparation into a single organization. It was not done to represent other same organization using data mining analysis. It was using the quantitative and qualitative method to explore data preparation. It began with a data collection approach to the analysis of data preparation. Although, it may be possible to generalize this paper.

The company set the principles of ethics, which was honored by the researcher. The company was informed that participating in this demonstration was voluntary and would not impact the company's brand. Ensuring anonymity, the paper removed some information that would be manipulation to favor the competitors [24]. Thus, the name of the organization is referred to as company A. Public information that could have damaged the company authenticity that could result in negative was removed.

4.1 Company description

Company A is one of the leading companies in producing steels. This company is situated in Alberton, where most of the production industries are built. It has a history of making several sheets of steel at a high rate. It increases the data in the dataset, not only proper data but also messy or dirty data. The company was selected due to its nature of producing a high number of products. Therefore, it was suitable for this research, which is dealing with data.

4.2 Data collection

Data collection is the method of gathering observations or collecting information using standard validated techniques. It is important to collect data to understand what can be done using it. Data collection consisted of two sources, which is primary and secondary data. Primary data refers to raw data collected. Secondary data is data that is already collected. Therefore, this paper selected secondary as company A already collected its data using sensors embedded in their machines into databases, repositories, and data warehouses.

The researcher extracted the dataset from the repository of company A based on the experience obtained through training in extracting data. This potential skill has helped the researcher to use data mining tool for preparing data. This was done during the period month of February and March 2021. Datasets were sent by company A to the researcher as the active participant in preparing data datasets due to the corona-virus pandemic. The datasets that were sent contained the machine, alarm data, and sensor data.

4.3 Data analysis

Data analysis is the process of systematically applying statistical and technique to evaluate data. According to [25], this type of research whereby data gathered is categorized into themes and sub-themes. Analysis helps data collected being reduced and simplified while at the same time producing results that may then measure using quantitative techniques. Moreover, the analysis provides the ability to the researcher to structure the qualitative data to satisfy the accomplishment of the paper objectives. The researcher installed a data mining tool as an "Add-on" to the Microsoft Excel spreadsheet. Microsoft Excel is a powerful tool for handling large data [26]. It consists of a grid with columns and rows that store data from resources of data. Data mining employed to arrange and remove inconsistencies that were on the datasets. Data mining was performed into a Microsoft Excel spreadsheet to prepare data for its readiness to be used for specific purposes. The resources were used in this paper are computer, Microsoft Excel spreadsheet, and data mining tool.

5. Results and discussion

This section describes the findings, and the overall discussion represents the datasets with data cleansing preparation. These three datasets were obtained from the data repository, and **Table 1** represents the excel files dataset before using the data mining tool. Machine data file contain 30 000 records in the dataset. It contains 10% missing values and 7 duplicate records. The alarm data file contains 45 000 records. It contains 25% missing values and 28 duplicated records in the dataset. Finally, the sensor data file contains 100 000 records in the dataset. It contains 45% missing values and 100 duplicated records. These files format was using Microsoft Excel as the technique to use datasets.

A data mining tool was used as the result of the analysis. **Table 2** shows the importance of using data mining in removing errors and inconsistencies in records. The data mining tool in **Table 2** has removed machine data records decreases from 30 000 to 26 993. There was no missing value found on the dataset, with no duplication records. Alarm data records decreased from 45 000 to 33 722, with no missing

Filename	No. of records	No. of fields	Missing value	Duplicated records
Machine data	30 000	1	10%	7
Alarm data	45 000	1	25%	28
Sensor data	100 000	1	45%	100

Table 1.
Raw data.

Filename	No. of records	No. of fields	Missing value	Duplicated records
Machine data	26 993	1	0%	0
Alarm data	33 722	1	0%	0
Sensor data	54 900	1	0%	0

Table 2.
Data mining uses.

values and duplicated records. Sensor data records decreased 100 000 to 54 900, with no missing values and duplicated records.

Missing values represent how efficient this tool in finding missing values of a file. Other features were whether this tool could find duplication, illegal values, merging the records and misspelling. Ease file format supported by these records and of use.

5.1 Discussion

This paper aims to investigate data cleansing in big data. Based on the available data cleansing methods discussed in the previous section, data cleansing for big data needs to be improvised and improved to cope with the massive amount of data. The traditional data cleaning method is important for developing the data cleaning framework for big data applications. In the review of Potter, this method only focused on solving data transformation challenges [13]. The Excel spreadsheet supports problems like duplicate record detection, and the user needs other approaches to deal with duplicate record detection problems [27].

Data mining can require manual and automatic procedures, but this approach focuses on duplication and missing elimination despite various data quality challenges in the dataset. Traditional data cleansing tools tend to solve only one data quality problem throughout the process and require human intervention to resolve data cleansing conflicts. In the big data era, the traditional data cleansing process is no longer acceptable as data needs to be cleansed and analyzed fast. The data is growing more complex as it may include structured data, semi-structured data, and unstructured data. The discussed methods focus only on structured data. However, existing methods have some limitations when working with dirty data. Data mining performs the computations of each stage as "local" in each Excel spreadsheet, and the data exchange is done at the stage boundaries by broadcast or hash partitioning.

6. Recommendations and conclusion

6.1 Recommendation

The chapter discusses the contribution of data mining cleaning on a dataset. This is achieved by discovering the errors and inconsistencies in the dataset and utilizing datasets stored in various resources. The authors discuss the importance of the management in organizations for attaching the vitality of data sourcing and strategic decision-making. The management must ensure that the correct, timely and accurate data is used in strategic decision-making to generate the ever-elusive competitive advantage. Furthermore, due to the key roles of the available data, big data has become a strategic resource. The data security required to be enhanced at all strategic decision-making levels to avoid unauthorized person (s) must be explored as future work.

6.2 Conclusion

Most organizations rely on data-driven decision making; therefore, the information system is closely related to business process management to leverage their processes for competitive advantage. Nowadays, the amount of data is constantly increasing, but the data quality is decreasing as much of the data collected is messy or dirty. There are various data cleansing approaches to solve this challenge, but data mining cleansing remains a tool to deal with the criteria of big data. Some of the approaches are not suitable for big data as there is a significant amount of data that needs to be processed simultaneously. Despite the availability of existing frameworks for data cleansing for big data, the value and veracity of the data are often disregarded while developing the approaches. Moreover, data mining is undeniably required to verify and validate the data before it can be subjected to an analysis process.

Acknowledgements

First, I would like to thank God for His blessing in completing this paper and my highest gratitude goes to my mentor for guarding me throughout this paper. Her patience on this paper was something I admired.

Also, thanks to seen and unseen hands that have given me direct and indirect help to finish this paper. Finally, thanks to my family who keeps encouraging through difficult time. Even if it was not fashionable to do so.

Declarations

I, Mawande Sikibi, hereby declare that this paper is wholly my work and has not been submitted anywhere else for academic credit either by myself or another person.

Author details

Mawande Sikibi
University of Johannesburg, South Africa

*Address all correspondence to: mawandesikibi@gmail.com

References

[1] Adelman-McCarthy JK. VizieR Online Data Catalog: The SDSS Photometric Catalog, Release 8 (Adelman-McCarthy+, 2011). VizieR Online Data Catalog. 2011 Sep:II-306.

[2] Buhrmester M, Kwang T, Gosling SD. Amazon's Mechanical Turk: A new source of inexpensive, yet high-quality data?.

[3] Cai L, Zhu Y. The challenges of data quality and data quality assessment in the big data era. Data science journal. 2015 May 22;14.

[4] Collis J, Hussey R. Business research: A practical guide for undergraduate and postgraduate students. Macmillan International Higher Education; 2013 Nov 29.

[5] Cong G, Fan W, Geerts F, Jia X, Ma S. Improving Data Quality: Consistency and Accuracy. InVLDB 2007 Sep 23 (Vol. 7, pp. 315-326).

[6] Gschwandtner T, Aigner W, Miksch S, Gärtner J, Kriglstein S, Pohl M, Suchy N. TimeCleanser: A visual analytics approach for data cleansing of time-oriented data. InProceedings of the 14th international conference on knowledge technologies and data-driven business 2014 Sep 16 (pp. 1-8).

[7] Hamilton WL, Ying R, Leskovec J. Inductive representation learning on large graphs. arXiv preprint arXiv:1706.02216. 2017 Jun 7.

[8] Hellerstein JM. Quantitative data cleaning for large databases. United Nations Economic Commission for Europe (UNECE). 2008 Feb 27;25.

[9] Jones S, Pryor G, Whyte A. How to Develop Research Data Management Services-a guide for HEIs.

[10] Kandel S, Heer J, Plaisant C, Kennedy J, Van Ham F, Riche NH, Weaver C, Lee B, Brodbeck D, Buono P. Research directions in data wrangling: Visualizations and transformations for usable and credible data. Information Visualization. 2011 Oct;10(4):271-88.

[11] Kusiak A. Data mining: manufacturing and service applications. International Journal of Production Research. 2006 Sep 15;44(18-19): 4175-91.

[12] Li D, Wang S, Li D. Spatial data mining. Berlin, Heidelberg: Springer Berlin Heidelberg; 2015.

[13] Liao X, Lochhead P, Nishihara R, Morikawa T, Kuchiba A, Yamauchi M, Imamura Y, Qian ZR, Baba Y, Shima K, Sun R. Aspirin use, tumor PIK3CA mutation, and colorectal-cancer survival. New England Journal of Medicine. 2012 Oct 25;367(17): 1596-606.

[14] Mathew PS, Pillai AS. Big Data solutions in Healthcare: Problems and perspectives. In2015 International Conference on Innovations in Information, Embedded and Communication Systems (ICIIECS) 2015 Mar 19 (pp. 1-6). IEEE.

[15] Moore CA, McCabe ER. Utility of Population-based Birth Defects Surveillance for Monitoring the Health of Infants and as a Foundation for Etiologic Research. Birth defects research. Part A, Clinical and molecular teratology. 2015 Nov;103(11):895.

[16] Olson DL, Delen D. Advanced data mining techniques. Springer Science & Business Media; 2008.

[17] Polyzotis N, Zinkevich M, Roy S, Breck E, Whang S. Data validation for machine learning. Proceedings of

Machine Learning and Systems. 2019 Apr 15;1:334-47.

[18] Rahm E, Do HH. Data cleaning: Problems and current approaches. IEEE Data Eng. Bull. 2000 Dec;23(4):3-13.

[19] Rajkumar SV. Multiple myeloma: 2012 update on diagnosis, risk-stratification, and management. American journal of hematology. 2012 Jan;87(1):78-88.

[20] Raman V, Hellerstein JM. Potter's wheel: An interactive data cleaning system. InVLDB 2001 Sep 11 (Vol. 1, pp. 381-390).

[21] Ridzuan F, Zainon WM. A review on data cleansing methods for big data. Procedia Computer Science. 2019 Jan 1;161:731-8.

[22] Van den Broeck J, Cunningham SA, Eeckels R, Herbst K. Data cleaning: detecting, diagnosing, and editing data abnormalities. PLoS Med. 2005 Sep 6;2(10):e267.

[23] Vandecruys O, Martens D, Baesens B, Mues C, De Backer M, Haesen R. Mining software repositories for comprehensible software fault prediction models. Journal of Systems and software. 2008 May 1;81(5):823-39.

[24] Wang L, Jacques SL, Zheng L. MCML—Monte Carlo modeling of light transport in multi-layered tissues. Computer methods and programs in biomedicine. 1995 Jul 1;47(2):131-46.

[25] Wirth R, Hipp J. CRISP-DM: Towards a standard process model for data mining. InProceedings of the 4th international conference on the practical applications of knowledge discovery and data mining 2000 Apr 11 (Vol. 1). London, UK: Springer-Verlag.

[26] Wu X, Zhu X, Wu GQ, Ding W. Data mining with big data. IEEE transactions on knowledge and data engineering. 2013 Jun 26;26(1):97-107.

[27] Zhang S, Zhang C, Yang Q. Data preparation for data mining. Applied artificial intelligence. 2003 May 1;17(5-6):375-81.

Chapter 3

Privacy Preserving Data Mining

Esma Ergüner Özkoç

Abstract

Data mining techniques provide benefits in many areas such as medicine, sports, marketing, signal processing as well as data and network security. However, although data mining techniques used in security subjects such as intrusion detection, biometric authentication, fraud and malware classification, "privacy" has become a serious problem, especially in data mining applications that involve the collection and sharing of personal data. For these reasons, the problem of protecting privacy in the context of data mining differs from traditional data privacy protection, as data mining can act as both a friend and foe. Chapter covers the previously developed privacy preserving data mining techniques in two parts: (i) techniques proposed for input data that will be subject to data mining and (ii) techniques suggested for processed data (output of the data mining algorithms). Also presents attacks against the privacy of data mining applications. The chapter conclude with a discussion of next-generation privacy-preserving data mining applications at both the individual and organizational levels.

Keywords: privacy preserving data mining, data privacy, PPDM methods, privacy attacks, Anonymization

1. Introduction

Especially with the 2019 pandemic, in today's world where business and education life is done electronically over the internet, fast and voluminous data sharing is made with the undeniable effect of social media and unfortunately technology works against privacy. The rapid widespread use of data mining techniques in areas such as medicine, sports, marketing, signal processing has also increased the interest in privacy. The important point here is to define the boundaries of the concept of privacy and to provide a clear definition. Individuals define privacy with the phrase "keep information about me from being available to others". However, when it comes to using these personal data in a study that is considered to be well intentioned, individuals are not disturbed by this situation and do not think that their privacy is violated [1]. What is missed here is the difficulty of preventing abuse once the information is released.

Personal data is information that relates to an identified or identifiable individual. This concept consists of the components that the data pertain to a person and that this person can also be identified. Personal data is a concept that belongs to the "ego" and is handled in a wide range from names to preferences, feelings and thoughts. An identifiable person is someone who can be identified directly or indirectly, in particular by reference to an identification number or one or more factors specific to their physical, physiological, mental, economic, cultural or social identity. For this reason, the loss of the individual's control authority over these data brings about the loss of the individual's freedom, autonomy, privacy, in short, the property of being me. The main way to ensure the use of these data without harming the privacy of individuals is to remove the identifiability of the person.

Data analysis methods, including data mining, commodify data and turn it into economic value. Apart from the ethical debates about this, it is an undeniable fact that the digital environment increases the risk of losing control of all information about one's own intellectual, emotional and situational, in short, losing its autonomy and violating the informational privacy area. The main dilemma here is; the freedom in the flow of information provided by technology, the interest relationships it provides and the benefit provided by the information source is the control power required by the concept of being an individual [2].

In addition, legal regulations aiming to protect personal data are made by governments, including for what purpose (historical, statistical, commercial, scientific) data is used, how it is collected and how it should be stored. For example, the US HIPAA rules aim to protect individually identifiable health information. These are information that is a subset of health information, including demographic information collected from an individual [3]. In the EC95/46 [4] directive, the European parliament and of the council allow the use of personal data in the case of (i) if the data subject has explicitly given his permission, or (ii) the need for a result requested by the individual. This also applies to corporate privacy issues. Privacy concerns bring corporate privacy concerns with them. However, corporate privacy and individual privacy issues are not much different from each other. The disclosure of information about an organization can be considered a potential privacy breach. In this case, it involves both views to generalize to disclosure of information about a subset of data.

The point to note here is that while focusing on the disclosure of data subjects, the secrets of the data providers' organization should also be taken into account. For example, considering that data mining studies were carried out on student data of more than one university in an academic study. Although the methods used protect the privacy of the student, certain information that is specific to the university and they want to keep may be revealed. Although the personal data owned by the organizations are secured by contracts and legal regulations, information about a subset of the combined data set may reveal the identity of the data subject. The organization that owns the data set must be involved in a distributed data mining process as long as it can prevent the disclosure of the data subjects it provides and its own trade secrets.

In the literature, solutions that take data privacy into account have been proposed in data mining. A solution that ensures that no individual data is exposed can still publish information that describes the collection as a whole. This type of corporate information is often the purpose of data mining, but some results can be identified, various data hiding and suppression techniques have been developed to ensure that the data are not individually identified.

The concept of privacy can be examined under three headings as "physical–physical, mental-communicative and data privacy [5]. The main subject in this study is data privacy.

1.1 Data privacy

Data privacy can be defined as the protection of real persons, institutions and organizations (Data Subject) that need to be protected in accordance with the law and ethical rules during the life cycle of data (collecting data, processing and analyzing data, publishing and sharing data, preserving data, re-use data) [6]. In this process, for what purpose the data will be processed, with whom it will be shared, where it will be transferred, and being able to be controlled by the data subject at a transparent and controllable level are important requirements of data privacy. On the other hand, there is no exact definition of privacy, the definition can be made specific to the application.

Data controllers who need to take privacy precautions in order to prevent data breaches are assumed to be reliable and have legal obligations; stores and uses the data collected with digital applications using appropriate methods, and shares them by anonymizing when necessary. Collected data are classified into four groups [7];

- Identifiers (ID): It contains information that uniquely and directly identifies individuals such as full name and social security number.

- Quasi-identifiers (QID): Identifiers that, combined with external data, lead to the indirect identification of an individual. These attributes are non-unique data such as gender, age, and postal code.

- Sensitive attributes (SA): It contains data that is private and sensitive to individuals, such as sickness and salary.

- Insensitive attributes: It contains general and non-risky data that are not covered by other attributes.

1.2 Privacy metrics

It is not sufficient to measure privacy with a single metric because different definitions can be made for different applications and multiple parameters must be evaluated for this purpose. It is possible to examine the proposed metrics for PPDMs [8, 9] as privacy level metric and data quality metric, depending on which aspect of privacy is measured. While evaluating these metrics, they can be measured in two subgroups to evaluate the level of privacy/data quality on the input data (data criteria) and data mining results (result criteria). How secure the data is in terms of disclosure is measured by the level of privacy metrics [10]:

Bounded knowledge: The purpose here is to restrict the data with certain rules and prevent the disclosure of the information that should remain confidential. It can be transformed into limited data by adding noise to the data or by generalizing the data.

Need to know: With this metric, keeping unnecessary data away from the system prevents privacy data that will arise. It also ensures that access control (access reason and access authorization) to data.

Protected from disclosure: In order to keep the confidential data that may come out as a result of data mining, some operations (such as checking the queries) can be done on the results to provide privacy. Using the classification method to prevent the disclosure of data, which is one of the criteria for ensuring privacy, is one of the effective methods [11].

Data quality metrics: It quantifies the loss of information/benefit, and the complexity criteria that measure the efficiency and scalability of different techniques are evaluated within this scope.

2. Data mining with privacy

Privacy Protected Data Mining (PPDM) techniques have been developed to allow the extraction of information from data sets while preventing the disclosure of data subjects' identities or sensitive information. In addition, PPDM allows more than one researcher to collaborate on a dataset [11, 12]. Also PPDM can be defined as performing data mining on data sets to be obtained from databases containing sensitive and confidential information in a multilateral environment without disclosing the data of each party to other parties [13].

In order to protect privacy in data mining, statistical and cryptographic based approaches have been proposed. The vast majority of these approaches operate on original data to protect privacy. This is referred to as the natural trade-off between data quality and privacy level.

PPDM methods are being studied on to perform effective data mining by guaranteeing a certain level of privacy. Several different taxonomies have been proposed for these methods. In the literature, based on data life cycle stages (data collection, data publishing, data distribution and output of data mining) [10] or they are classified based on the method used (Anonymization based, Perturbation based, Randomization based, Condensation based and Cryptography based) [14].

In this study, PPDM approaches are examined with a simple taxonomy as methods applied to input data and processed data (output information) that is subject to data mining.

2.1 Methods applied to input Data

This section includes the methods suggested for collecting, cleaning, integration, selection and transformation phases of input data that will be subject to data mining.

Although it varies according to the application used or the state of trust to the institution collecting the data, it is recommended that the original values not be stored and used only in the conversion process in order to prevent disclosure of privacy. For example, the data collected with sensors, which are now widely used with internet of things, can be transformed at the stage it collects, randomizing the obtained values and transforming the raw data before being used in data mining.

In this section, data perturbation, randomization, suppression, data swapping, anonymity, cryptography and differential privacy methods are discussed.

2.1.1 Data perturbation

The creation of data resistant to privacy attacks can be done by perturbation significantly preserving the statistical integrity of the data [15, 16]. Randomization of the original data is widely used in data perturbation [17–19]. Another approach is the Microaggregation method [20].

In the randomization method, noise signals are added to the data with a known statistical distribution, so when data mining methods are applied, the original data distribution can be reconstructed without accessing the original data. For this, data providers first randomize their data and then transmit them to the data recipient. Then, receiving this random data, the data receiver calculates the distribution using distribution reconstruction methods.

During the data collection phase, it can be calculated independently for each data, and after the original distribution is reconstructed, the statistical properties of the data are preserved. For example; the result of the randomization of A with B is C (C = A + B) if A be the original data distribution, and B, a publicly known noise distribution independent of A. Then, A may be reconstructed with "A= C− B". However, this reconstruction process may not be successful if B has a large variance and C's sample size is not large enough. As a solution, approaches that implement the Bayes [21], or EM [22] formula can be used. While the randomization method limits data usage to the distribution of C, it requires a lot of noise to hide outliers. Because in this approach, outliers are more vulnerable to attacks when compared to values in denser regions in the data. Although this reduces the use of the data for mining purposes, it may be necessary to add too much noise to all records in the data that would result in loss of information, in order to prevent it [7].

Randomly generated values can be added to the original data with an additive or multiplicative method [23]. The aim is to ensure that noise added to individual records for privacy is non-extractable. Multiplicative Noise is more efficient than the Additive Noise method because it is more difficult to predict the original values.

With Microaggregation method, all records in the data set are first arranged in a meaningful order and then the whole set is divided into a certain number of subsets. Then, by taking the average of the value of each subset of the specified attribute, the value of that attribute of the subset is replaced with the average value. Thus, the average value of that attribute for the entire data set will not change.

Since data perturbation approaches have a negative impact on data utility and are not resistant to attacks, they are often not preferred in utility-based data models.

2.1.2 Suppression

Data Suppression technique is a technique that tries to prevent the disclosure of confidential information by replacing some values with a special value. In some cases, it is the process of deleting cell values or the entire record [24]. In this way, confidential data can be changed, rounded, generalized or mixed and made available in data mining applications [25].

An example of Suppression may be changing the age attribute in records from 28 to 35, city attribute from Glasgow to Edinburgh, or generalizing the age attribute from 28 to 25–30, and Glasgow data as Scotland. Using these methods in big data can reduce data quality and change general statistics, this may result in data becoming unusable [26]. Another problem is that information is deliberately distorted to suppression. Data providers can obtain artificial inferences that are inaccurate and serve a purpose with the reported values [27].

On the other hand, suppression should not be used when data mining requires full access to sensitive values. For sensitive information in a record, the method of limiting the identity link of a record may be preferred instead.

2.1.3 Data swapping

A technique tries to prevent the disclosure of private information by swapping values between different records.

Data swapping can be explained as each data provider scrambling data by exchanging their data with other data providers, especially in cases where there are more than one data provider. The advantage of the technique is that the data does not affect the sub-order sums, thus allowing accurate and complete collective calculations.

With this technique, as the result of data exchanges, private data can be easily exposed in the system, for this reason it is recommended to use only in safe environments. It can be used in conjunction with other methods such as k-anonymity without violating privacy definitions.

2.1.4 Cryptography

Cryptography is a technique that converts plain text to cipher text using various encryption algorithms to encode messages in a way that cannot be read. It is a method of storing and transmitting data in specific form using cryptography techniques so that only intended persons can read and process it.

In data mining applications, cryptography-based techniques are used to protect privacy during data collection and data storage [25, 28], and guarantee a very high level of data privacy [23]. Encryption is generally costly due to time and computational complexity. Hence, as the volume of data increases, the time to process on encrypted data increases and creates a potential barrier to real-time analysis [29].

Secure multiparty computing (SMC) is a special encryption protocol where, when there is more than one participating party, the interested parties learn nothing but results [30, 31]. The SMC calculation must be done carefully so that it does not reveal sensitive data, but the calculated result can enable the parties to estimate the value of sensitive data.

2.1.5 Group-based anonymization

Many privacy conversions are for creating groups between anonymous records that are converted in a group-specific manner. A number of techniques have been proposed for group anonymity in different studies, such as k-anonymity, l-diversity, and t-proximity methods. The comparison of group anonymity methods is given in **Table 1**.

2.1.5.1 k-anonymity

The k-anonymity method proposed by Samarati and Sweeney in the anonymization of data is a method of providing privacy that protects the identity of the data subject most commonly used in the publication of data [32].

The method ensures that after removing the ID attributes from the table, the QID values of at least k records in the table to be published are the same.

Since the QID attributes of each record in the table published by this method are the same as the other k-1 records, it is aimed to prevent identity disclosure.

To reduce the level of detail of the data representation, some attributes can be replaced with more general values (data swapping), some data points can be eliminated, or descriptive data can be deleted (suppression). However, while k-Anonymity provides protection against attacks on the disclosure of identities, it does not protect against attacks on disclosure of attributes. It is also more convenient to use for individual data rather than directly applying it to restrict data mining results that protect privacy. Besides, k-anonymity fully protects the privacy of users when it comes to the homogeneity of sensitive values in the data. Providing optimum k-anonymity is a problem in the NP-Hard class and approximate solutions have been proposed to avoid calculation difficulties [33].

In the literature, different studies such as k-neighbor anonymity, k-degree anonymity, cotomorphism anonymity, k-candidate anonymity and l-grouping derived

Method	Based on	Vulnerability under	Strong against
k-anonymity	Sensitive data disclosure	Homogeneity attack	record linkage only
l-diversity	Semantic similarity of sensitive data	Skewness attack	record linkage and attribute linkage
t-closeness	Distance measures	Attribute linkage attack	probabilistic attack and attribute linkage

Table 1.
Group based anonymity methods.

from the k-anonymity approach have been proposed according to the structural features of the data.

2.1.5.2 l-diversity

The l-diversity approach was proposed by Ashwin Machanavijjhala in 2007 to address the weaknesses (homogeneity attack) of the k-anonymity model [34].

This method aims to prevent the disclosure of confidential information indirectly by ensuring that each QID group has at least l well-represented sensitive value.

L-diversity only guarantees the diversity of sensitive features within each QID group, but the problem that different values may belong to the same category is not solved.

In other words, it is not resistant to attacks based on semantic similarity between values.

2.1.5.3 t-closeness

In order to balance the semantic similarities of SA attributes within each QID group, it has been proposed to solve the limitations of the l-diversity approach by guaranteeing t-closeness to each other [35].

Accordingly, in t-closeness method, the distance of the distribution of sensitive attributes in any equivalence class to the distribution of the attributes in the whole table will not exceed a threshold value (t). While the t-closeness approach provides protection against disclosure of attributes, it cannot protect against disclosure of identities. In addition, it limits the usefulness of the information disclosed however, by setting the t-threshold in applications, it can exchange benefit and privacy.

In the protection of privacy, t-proximity and k-anonymity methods are used together to protect against attacks on identity disclosure and quality [36].

2.2 Methods applied to processed Data

The outputs of data mining algorithms can disclose information without open access to the original data set. Sensitive information can be accessed through studies on the results. For this reason, data mining output must also protect privacy.

2.2.1 Query auditing and inference control

This method is examined as query inference control and query auditing. In the query inference control, the input data or the output of the query is controlled. In t Query auditing, the queries made on the outputs obtained by data mining are audited. If the audited query enables the disclosure of confidential data, the query request is denied. Although it limits data mining, it plays an active role in ensuring privacy. Query auditing can be done online or offline. Since queries and query results are already known in offline control, it is evaluated whether the results violate privacy. In online auditing, since the queries are not known, privacy metrics are carried out simultaneously during the execution of the query. This method is examined within the scope of statistical database security.

2.2.2 Differential privacy

k-anonymity, l-diversity and t-closeness approaches are holistic approaches that try to protect the whole data privacy. In some cases, there is a need to protect the privacy of data at the record level. For this reason, differential privacy approach has been proposed by Dwork to protect the privacy of database query results [37].

With this model, the attacks that may occur between sending database queries and responding to the query are targeted. Failure to distinguish from which database the answer of the same query, made in more than one database, is returned will prevent the disclosure of the existence of a single record between databases.

In addition, when querying output data, it can be ensured that the query results obtain approximate values with the database approach technique. Also, it is recommended to keep the data in the system mixed during the execution of queries, just like the data collection phases to protect data privacy.

2.2.3 Association rule hiding

In data mining, it is one of the most frequently used methods of Association Rules to reveal the nature of interesting associations between binary variables. During data mining, some rules may explicitly disclose private information about the data subject (individual or group).

Unnecessary and information-leaking rules may occur in some relationships. The aim of the Association rule hiding technique first proposed by Atallah [38] is to protect privacy by hiding all sensitive rules. The weakness with this technique is that a significant number of insensitive rules can be hidden incorrectly [39].

3. Attacks against privacy

In this section, the common types of attacks that lead to the development of the methods given above and lead to privacy violations are summarized [6].

3.1 Semantic similarity attacks

Attacks that are made by making use of the intuitive similarity of sensitive attribute values within anonymous groups.

In this case, it is not sufficient for the sensitive attribute values to be different from each other in terms of protecting privacy [40]. This attack can be prevented by calculating the similarities of sensitive attributes in the same anonymous group and by providing solutions to include similar sensitive attribute values in different groups.

3.2 Background knowledge attacks

Background knowledge is non-sensitive information that can be obtained from data published by different organizations, social networks and media even by using social engineering methods. Background knowledge obtained by attacker's causes privacy attacks and breaches.

Data subject's privacy violation occurs as a result of associating background knowledge with other records using data binding methods [41].

In addition, when information obtained from data owners through requests such as promotion, campaign, research, etc. is associated with background information, it is not even possible that it will not cause a violation of privacy.

3.3 Homogeneity attacks

In cases where all or most of the sensitive attributes in the groups included in the anonymous tables are similar, the privacy of data owners is at risk of violation.

In order to prevent homogeneity attacks, it is necessary to prevent similar sensitive attributes within the groups in the anonymous table from being in the same group or to reproduce heterogeneous records by diluting the homogeneous attributes with the record duplication approach [34].

3.4 Skewness attacks

The statistical distribution of sensitive attribute values in published or shared anonymous data sets can lead to the success of skewness attacks against privacy. The distortion in the general distribution of sensitive attributes occurs when these values are too dominant and anonymous data sets become vulnerable [35].

3.5 De-Finetti attacks

It has been shown that with theoretical and experimental methods, interchangeability concepts and inferences about privacy can be made with Definetti's theorem [42]. The fact that the people who want to carry out this attack do not need extensive background knowledge makes this attack attractive. An attacker can perform an attack using machine-learning techniques on non-sensitive attributes in the dataset.

3.6 Minimality attacks

The fact that the information about which data anonymization algorithm is used in the data mining application is public is also considered as a privacy vulnerability [43]. It is based on the principle that changes on data should remain at minimum level in anonymization processes and should not be overly anonymized.

3.7 Temporal attack

Publicly declaring previously published generalized data over time causes this attack. For this reason, previously published tables should be used and new records that may cause data disclosure should not be shared [44].

4. Discussion

The fact that the digitalization process has become mandatory all over the world with Covid-19 pandemic has accelerated the data flow. It has become even more important to collect the necessary data, analyze it correctly and reveal reliable information. This situation has triggered the use of data mining methods to increase productivity and provide high quality products/services in almost all sectors. While applying data mining methods, it is obvious that if privacy is not taken into consideration during the data life cycle, irreversible damages will occur for individuals/institutions and organizations.

In order to increase the access and benefits of data mining technology, before applying PPDM techniques, "privacy" should be defined precisely, measurement metrics should be determined and the results obtained should be evaluated with these metrics. For this reason, this study primarily focused on the definition of privacy. The term privacy is quite extensive and does not have a standard definition. It is quite challenging in measuring privacy, as there is no standard privacy definition. Some measurement metrics are mentioned in this chapter, but metrics are usually determined by application. The lack of a standard privacy measurement metric also make challenging the comparison and evaluation of the developed PPDM techniques.

In the age of digital and online business, privacy protection needs to be done at the individual and organizational levels. Privacy protection at the individual level depends on person who is influenced by religious beliefs, community norms and culture. For this reason, the concept of personalized privacy, which allows individuals to have a certain level of control over their data, has been proposed. However, it has been observed that there are difficulties in implementing personalized privacy, as people think that compromising their privacy for applications they think is well-intentioned will not damage. Therefore, in the context of personalized privacy, new solutions are required for the trade-off between privacy and utility.

To effectively protect organizational level data privacy [7]; Policy makers in organizations should support privacy-enhancing technical architectures/models to securely collect, analyze and share data. Laws, regulations and fundamental principles regarding privacy should be analyzed by organizations. It is necessary for organizations to include the data owners in their assessment of privacy and security practices. Data owners should involve the whole process about what data is collected, how it is analyzed and for what purpose it is used. In addition, they should have the right to correct personal data in order to avoid negative consequences of incorrect data. Organizations should employ data privacy analysts, data security scientists, and data privacy architects who can develop data mining applications securely.

From a technical point of view, methods that protect confidentiality in data analytics are still in their infancy. Although studies continue by different scientific communities such as cryptography, database management and data mining, an interdisciplinary study should be conducted on PPDM. For example, the difficulties encountered in this process should also be addressed from a legal perspective. Thus, a better roadmap for next-generation privacy-preserving data mining design can be developed by academic researchers and industrial practitioners.

5. Conclusion

Businesses and even governments collect data through many digital platforms (social media, e-health, e-commerce, entertainment, e-government etc.) they use to serve their customers/citizens. The data collected can be sensitive data and this data can be stored, analyzed and, in good probability, anonymized and shared with others. In studies where data is used at any stage of the life cycle, regardless of the purpose, it is necessary to explain a privacy permission and the reason why the data should be accessed. Privacy Preserving Data Mining (PPDM) techniques are being developed to allow information to be extracted from data without disclosing sensitive information.

There is no single optimal PPDM technique for any stage of the data lifecycle. The PPDM technique to be applied varies according to the application requirements, such as the desired privacy level, data size and volume, tolerable information loss level, transaction complexity, etc. Because different application areas have different rules, assumptions and requirements regarding privacy.

In this chapter, the previously proposed PPDM techniques are examined in two sections. First section includes the methods suggested for collecting, cleaning, integration, selection and transformation phases of input data that will be subject to data mining and second section covers methods applied to processed data. Finally, attacks against the privacy of data mining applications are given in this chapter.

Author details

Esma Ergüner Özkoç
Başkent University, Ankara, Turkey

*Address all correspondence to: eeozkoc@baskent.edu.tr

References

[1] Clifton C, Kantarcioglu M, Vaidya J, Defining privacy for data mining. In National science foundation workshop on next generation data mining. 2002; Vol. 1, No. 26, p. 1

[2] İzgi M. C, The concept of privacy in the context of personal health data. Türkiye Biyoetik Dergisi, 2014. (S 1), 1

[3] Centers for Disease Control and Prevention. HIPAA privacy rule and public health. Guidance from CDC and the US Department of Health and Human Services. MMWR: Morbidity and mortality weekly report, 200352(Suppl 1), 1-17.

[4] Data P, Directive 95/46/EC of the European parliament and of the council on the protection of individuals with regard to the processing of personal data and on the free movement of such data. Official Journal L, 1995; 281(23/11), 0031-0050.

[5] Belsey A, Chadwick, R. Ethical issues in journalism and the media. Routledge. (Eds.) 2002

[6] Vural Y, Veri Mahremiyeti: Saldırılar, Korunma Ve Yeni Bir Çözüm Önerisi. Uluslararası Bilgi Güvenliği Mühendisliği Dergisi, 4(2), 21-34.

[7] Pramanik M. I, Lau R. Y, Hossain M. S, Rahoman M. M, Debnath S. K, Rashed, M. G., Uddin M. Z.,. Privacy preserving big data analytics: A critical analysis of state-of-the-art. Wiley Interdisciplinary Reviews: Data Mining and Knowledge Discovery, 2021; 11(1), e1387.

[8] Bertino E, Lin D, Jiang W, A survey of quantification of privacy preserving data mining algorithms, in Privacy-Preserving Data Mining. New York, NY, USA: Springer, 2008, pp. 183-205.

[9] Dua S, Du X, Data Mining and Machine Learning in Cybersecurity. Boca Raton, FL, USA: CRC Press, 2011.

[10] Mendes R, Vilela J. P, Privacy-preserving data mining: methods, metrics, and applications. IEEE Access, 2017; 5, 10562-10582.

[11] Vaidya J, Clifton C, Privacy-preserving data mining: Why, how, and when. IEEE Security & Privacy, 2004; 2(6), 19-27.

[12] Nayak G, Devi S, A survey on privacy preserving data mining: approaches and techniques. International Journal of Engineering Science and Technology, 2011; 3(3), 2127-2133.

[13] Lindell Y, Pinkas B, Privacy Preserving Data Mining, In: Proceedings of the 20th Annual International Cryptology Conference, 2000; California, USA, 36- 53

[14] Rathod S, Patel D, Survey on Privacy Preserving Data Mining Techniques. International Journal of Engineering Research & Technology (IJERT) 2020; Vol. 9 Issue 06

[15] Hong T. P, Yang K. T, Lin C. W, Wang S. L, Evolutionary privacy-preserving data mining. In: Proceedings of the World Automation Congress 2010; (pp. 1-7). IEEE.

[16] Qi X, Zong M, An overview of privacy preserving data mining. Procedia Environmental Sciences, 2011; 12, 1341-1347

[17] Muralidhar K, Sarathy R, A theoretical basis for perturbation methods. Statistics and Computing, 2003; 13(4), 329-335

[18] Evfimievski A, Randomization in privacy preserving data mining. ACM

Sigkdd Explorations Newsletter, 2002; 4(2), 43-48

[19] Kargupta H, Datta S, Wang Q, Sivakumar K, On the privacy preserving properties of random data perturbation techniques. In: Proceedings of the Third IEEE international conference on data mining 2003; (pp. 99-106). IEEE.

[20] Domingo-Ferrer J, Torra V, Ordinal, continuous and heterogeneous k-anonymity through microaggregation. Data Mining and Knowledge Discovery, 2005; 11(2), 195-212

[21] Agrawal R, Srikant R, Privacy-preserving data mining. In Proceedings of the 2000 ACM SIGMOD international conference on Management of data 2000; (pp. 439-450).

[22] Agrawal D, Aggarwal C. C, On the design and quantification of privacy preserving data mining algorithms. In Proceedings of the Twen tieth ACM SIGMOD-SIGACT-SIGART Symposium on Principles of Database Systems 2001; pp. 247-255.

[23] Niranjan A, Nitish A, Security in Data Mining-A Comprehensive Survey. Global Journal of Computer Science and Technology 2017

[24] Oliveira S, Zaiane O, Data perturbation by rotation for privacy-preserving clustering, Technical Report 2004.

[25] Verykios V. S, Bertino E, Fovino I. N, Provenza L. P, Saygin Y, Theodoridis Y, State-of-the-art in privacy preserving data min ing. ACM SIGMOD Record, 2004; 33, 50-57.

[26] Aggarwal C. C, On randomization, public information and the curse of dimensionality. In: Proceedings of the IEEE 23rd International Conference on Data Engineering; Istanbul, Turkey, 2007, pp. 136-145.

[27] Zhu D, Li X. B, Wu S, Identity disclosure protection: A data reconstruction approach for privacy-preserving data mining. Decision Support Systems, 2009; 48, 133-140.

[28] Yang Y, Zheng X, Guo W, Liu X, Chang V, Privacy-preserving smart IoT-based healthcare big data storage and self-adaptive access control system. Information Sciences, 2019; 479, 567-592.

[29] Lu, R., Zhu, H., Liu, X., Liu, J. K., & Shao, J. (2014). Toward efficient and privacy-preserving computing in big data era. IEEE Network, 28, 46-50.

[30] Yao A. C, How to generate and exchange secrets. In: Proceedings of the 27th IEEE Symposium on Foundations of Computer Science, 1986; 162-167. IEEE

[31] Goldreich O, Micali S, Wigderson A, How to play any mental game - a completeness theorem for protocols with honest majority. In: Proceedings of the 19th ACM Symposium on the Theory of Computing, 1987; 218-229.

[32] Sweeney L, k-anonymity: A model for protecting privacy. International Journal of Uncertainty, Fuzziness and Knowledge-Based Systems, 2002; 10(05), 557-570.

[33] Samarati P, Sweeney L, Protecting privacy when disclosing information: k-anonymity and its enforcement through generalization and suppression, SRI International, Technical Report, 1998; SRI-CSL-98-04

[34] Machanavajjhala A, Gehrke J, Kifer D, Venkitasubramaniam M, ℓ-Diversity: Privacy beyond k-anonymity, In: Proceedings of the The 22nd International Conference on Data Engineering, 2006; Atlanta, USA,

[35] Li N, Li T, Venkatasubramanian S, t-Closeness: Privacy beyond

k-anonymity and ℓ-diversity, In: Proceedings of the International Conference on Data Engineering (ICDE), İstanbul, Turkey, 2007; 106-115

[36] Rubner Y, Tomasi C, Guibas L. J, The earth mover's distance as a metric for image retrieval. International journal of computer vision, 2000; 40(2), 99-121.

[37] Dwork C, Differential privacy: A survey of results. In: Proceedings of the International conference on theory and applications of models of computation Springer, Berlin, Heidelberg. 2008; (pp. 1-19).

[38] Atallah M, Bertino E, Elmagarmid A, Ibrahim M, Verykios V, Disclosure limitation of sensitive rules. In Knowledge and Data Engineering Exchange Workshop (KDEX'99), 1999; 25-32.

[39] Evfimievski A, Srikant R, Agrawal R, Gehrke J, Privacy preserving mining of association rules. In: Proceedings of the Eighth ACM SIGKDD International Conference on Knowledge Discovery and Data Mining, 2002; 217-228.

[40] Wang H, Han J, Wang J, Wang L, (l, e)- Diversity - A Privacy Preserving Model to Resist Semantic Similarity Attack, Journal of Computers, 2014; 59-64

[41] Chen B. C, LeFevre K, Ramakrishnan R, Privacy skyline: Privacy with multidimensional adversarial knowledge. University of Wisconsin-Madison Department of Computer Sciences. 2007

[42] Kifer D, Attacks on privacy and deFinetti's theorem", In: Proceedings of the ACM SIGMOD International Conference on Management of data, Rhode Island, ABD, 2009; 127-138, 2009

[43] Wong R. C. W, Fu A. W. C, Wang K, Pei J, Minimality attack in privacy preserving data publishing. In: Proceedings of the 33rd international conference on Very large data bases 2007; (pp. 543-554).

[44] Sanjita B. R, Nipunika A, Desai R, Privacy Preserving In Data Mining, Journal of Emerging Technologies and Innovative Research 2019; vol6 Issue 5

Chapter 4

Multilabel Classification Based on Graph Neural Networks

Wei-Cheng Ye and Jia-Ching Wang

Abstract

Typical Laplacian embedding focuses on building Laplacian matrices prior to minimizing weights of connected graph components. However, for multilabel problems, it is difficult to determine such Laplacian graphs owing to multiple relations between vertices. Unlike typical approaches that require precomputed Laplacian matrices, this chapter presents a new method for automatically constructing Laplacian graphs during Laplacian embedding. By using trace minimization techniques, the topology of the Laplacian graph can be learned from input data, subsequently creating robust Laplacian embedding and influencing graph convolutional networks. Experiments on different open datasets with clean data and Gaussian noise were carried out. The noise level ranged from 6% to 12% of the maximum value of each dataset. Eleven different multilabel classification algorithms were used as the baselines for comparison. To verify the performance, three evaluation metrics specific to multilabel learning are proposed because multilabel learning is much more complicated than traditional single-label settings; each sample can be associated with multiple labels. The experimental results show that the proposed method performed better than the baselines, even when the data were contaminated by noise. The findings indicate that the proposed method is reliably robust against noise.

Keywords: graph neural networks, multilabel classification, deep learning

1. Introduction

Traditional supervised learning deals with the analysis of single-label data, which means that samples are associated with a single label. However, in many real-world data mining applications, such as text classification [1, 2], scene classification [3, 4], crowd sensing/mining [5–11], and gene functional classification [12, 13], the samples are associated with more than one label. From this description, we understand that the challenge of the multilabel classification task is its potential output.

Basically, multilabel learning algorithms can be categorized into two different groups. 1) Problem transformation method. This method takes the multilabel problem and converts it into a single-label problem that can easily be classified using any classifier using the relationship between labels. 2) Adapted algorithm method. This method directly performs multilabel classification rather than transforming the problem into different subsets of problems, and most of these methods use the Euclidean distance between samples.

The main idea of this paper is to aggregate similar samples to obtain better results. To aggregate similar samples, we use the properties of graph neural networks (GNNs) [14]. The main contributions of this study are as follows:

- We propose a method that constructs a multilabel-based Laplacian graph such that each element in it represents the relationship between samples.
- We use similar samples with an aggregation approach that is not used in traditional multilabel learning methods.

The rest of this paper is arranged as follows. Section 2 shows the taxonomy of multilabel learning algorithms and describes their methods. Section 3 presents the details of our proposed method. Section 4 describes the multilabel datasets, evaluation metrics and experimental results, followed by the conclusions in Section 5.

2. Related work

2.1 Multilabel learning algorithms

In this section, we review multilabel learning algorithms. The algorithms that have been applied to multilabel learning over the last decade are not just those mentioned in this paper. **Figure 1** summarizes the algorithms detailed in the next section.

2.1.1 Problem transformation

Binary relevance (BR) is used to address a multilabel problem with a binary classifier, and its advantages are simplicity and efficiency, but correlation between labels is not considered. Classifier chains (CCs) are configured in a chain of binary classifiers where a classifier in the chain is based on the prediction of the previous classifier; their advantage is that they consider the relationship between labels but hence cannot be parallelized. Calibrated label ranking (CAL) performs ranking via the pairwise comparison of labels and has the advantage of considering the

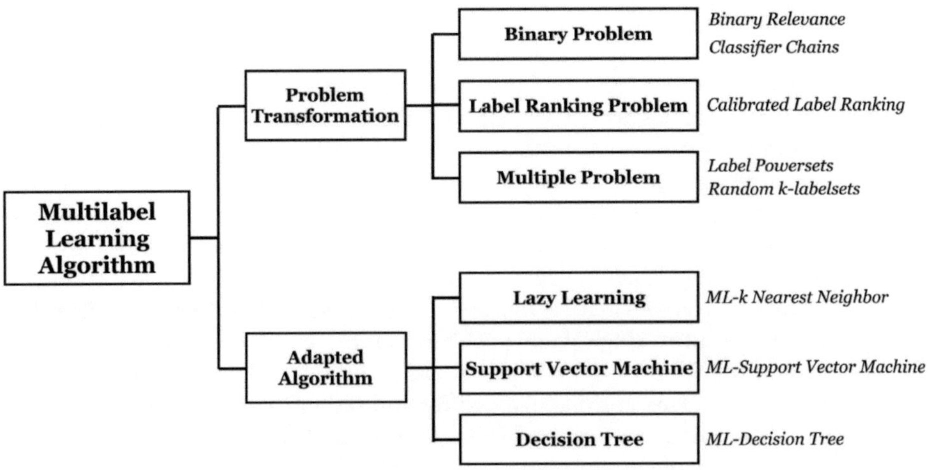

Figure 1.
Taxonomy of multilabel learning algorithms [15].

relationship (but only the pairwise relationship) between labels. Label powersets (LP) treat the situation when multiple labels belong to the same sample as a new label and have the advantage of considering the relationship between labels, but the time complexity grows exponentially with label sets. Random k-labelsets (RKL) are variants of LP models where each classifier is trained with a small random set of labels; their advantage is that they consider the relationship between labels, but they have a low accuracy rate if a worse label set combination is randomly selected.

2.1.2 Adapted algorithm

The multilabel k-nearest neighbor (MLkNN) method is derived from the traditional k-nearest neighbor algorithm. Each sample is identified with k nearest neighbors in the training set, and information is obtained from these identified neighbors. Multilabel support vector machine (ML-SVM) classification determines an optimal hyperplane that separates observations according to their labels. A multilabel decision tree (ML-DT) is constructed by building a decision tree, where each node corresponds to a set of samples in the data set.

2.2 Graph neural networks

GNNs were mentioned for the first time and further elaborated by [16]. The goal of a GNN is to learn a node's representation of the acquisition of its information by propagation. Currently, there are many deep learning tasks that need to process data with graph structures. Convolutional neural networks (CNNs) [17] have been successfully developed in the field of computer vision [18, 19] but are unable to process graph structured data [20]. The method used in this paper is called a graph convolutional network (GCN). A GCN can aggregate similar samples by propagating neighbor information, giving it the ability to infer, and there is no need to consider the sequence. GCNs have appeared in many top machine learning conferences and many applications across different tasks and domains, such as manifold learning [21, 22], computer vision [23–25], text classification [26, 27], hashing [28, 29], and hyperspectral image classification [30, 31].

3. The proposed method

This section presents the overall flow of our proposed method, as shown in **Figure 2**. The multilabel data matrix is first converted into a similarity matrix generated from a Laplacian graph. We call this a multilabel-based Laplacian graph and use this graph as inputs to the GCN model. Each node in the output layer predicts the probability of class membership for the label.

3.1 Multilabel-based Laplacian graph

This section presents the proposed method. Before this, let us describe some notational conventions. Matrices are written in boldface capital letters (e.g., $\mathbf{X}$). The transpose of a matrix is denoted as $\mathbf{X}^\top$. Vectors are written in boldface lowercase letters (e.g., $\mathbf{x}$). For a matrix $\mathbf{X} \in \mathbb{R}^{n \times m}$, the j-th column and the ij-th entry are denoted by $\mathbf{x}_j$ and x_{ij}, respectively. $\mathbf{I}$ denotes the identity matrix, $\|\cdot\|_2$ is the l_2-norm, and $\mathbf{1}$ denotes a column vector with all elements equal to ones.

Based on [32], we formally present our multilabel-based Laplacian graph. For a multilabel dataset, let $\mathbf{X} = [\mathbf{x}_1, \cdots, \mathbf{x}_n] \in \mathbb{R}^{n \times m}$ be the data matrix with n and m

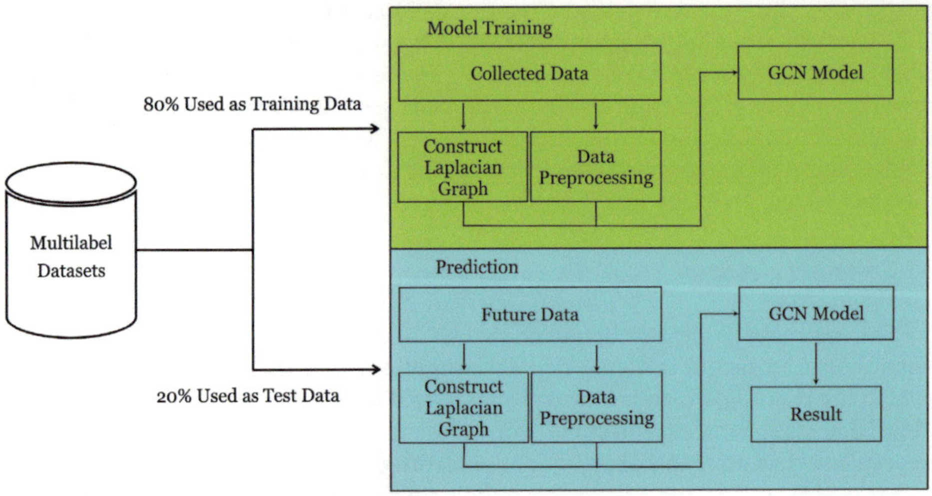

Figure 2.
An illustration of the work flow of the proposed method. Fully green color represents the training model; fully blue color represents the test model.

representing the number of samples and the dimensions, respectively. $\mathbf{S} \in \mathbb{R}^{n\times n}$ is the multilabel-based Laplacian graph, and we use a sparse representation method to construct this graph as follows:

$$\min_{\mathbf{S}} \sum_{i,j=1}^{n} \left\|\mathbf{x}_i - \mathbf{x}_j\right\|_2^2 S_{ij} + \beta \sum_{i=1}^{n} \|\mathbf{s}_i\|_2^2$$
$$\text{s.t. } \forall S_{ii} = 0,\ S_{ij} \geq 0,\ \mathbf{1}^T \mathbf{s}_i = 1. \tag{1}$$

We normalize $\mathbf{1}^T\mathbf{s}_i = 1$ which represents a sparse constraint on $\mathbf{S}$ because sparse representation is robust to noise [33], and β is an adjustable parameter. The second term is added to regularize the loss function.

3.2 Graph convolutional network

Based on [34], we fit the GCN used for single-label classification to multilabel classification. The GCN has been modified from a first-order Chebyshev approximation [35]. In order to create a multidimensional input, ChevNet convolution with an input vector $\mathbf{x}$ and a filter g_θ is formulated as follows:

$$\mathbf{x}\star g_\theta = \theta_0 \mathbf{x} - \theta_1 \mathbf{D}^{-\frac{1}{2}} \mathbf{A} \mathbf{D}^{-\frac{1}{2}} \mathbf{x}, \tag{2}$$

where $\star$ means the convolution operator, $\mathbf{A}$ is the adjacency matrix and $\mathbf{D}$ is the degree matrix. By using the single parameter $\theta = \theta_0 = -\theta_1$ to avoid overfitting, Eq. (2) can be rewritten as:

$$\mathbf{x}\star g_\theta = \theta\left(\mathbf{I}_n + \mathbf{D}^{-\frac{1}{2}} \mathbf{A} \mathbf{D}^{-\frac{1}{2}}\right)\mathbf{x}. \tag{3}$$

Repeated use of this graph convolution operation may cause serious problems such as vanishing gradients. Therefore, $\mathbf{I}_n + \mathbf{D}^{-\frac{1}{2}}\mathbf{A}\mathbf{D}^{-\frac{1}{2}}$ in Eq. (3) is modified to $\tilde{\mathbf{D}}^{-\frac{1}{2}}\tilde{\mathbf{A}}\tilde{\mathbf{D}}^{-\frac{1}{2}}$ with $\tilde{\mathbf{A}} = \mathbf{A} + \mathbf{I}_n$ and $\tilde{\mathbf{D}}_{ii} = \sum_j \tilde{\mathbf{A}}_{ij}$, finally giving a layerwise propagation rule to support multidimensional inputs as follows:

$$\mathbf{H}^{(l+1)} = \sigma\left(\tilde{\mathbf{D}}^{-\frac{1}{2}}\tilde{\mathbf{A}}\tilde{\mathbf{D}}^{-\frac{1}{2}}\mathbf{H}^{(l)}\mathbf{W}^{(l)}\right). \tag{4}$$

Here, $\mathbf{H}^{(l)}$ is the output of an activation function in the l-th layer of the GCN. $\mathbf{W}^{(l)}$ is a trainable weight matrix corresponding to the l-th layer of GCN. $\mathbf{H}^{(0)}$ is the data matrix. $\sigma(\cdot)$ denotes a specific activation function such as a sigmoid activation function.

This paper considers only a two-layer GCN model as the proposed method, and we modify Eq. (4) by placing the adjacent matrix into a multilabel-based Laplacian graph to obtain the formula of the two-layer GCN method proposed in this paper as follows:

$$\begin{aligned}\mathbf{H}^{(1)} &= \sigma\left(\hat{\mathbf{D}}^{-\frac{1}{2}}\hat{\mathbf{S}}\hat{\mathbf{D}}^{-\frac{1}{2}}\mathbf{H}^{(0)}\mathbf{W}^{(0)}\right)\\ \mathbf{H}^{(2)} &= \sigma\left(\hat{\mathbf{D}}^{-\frac{1}{2}}\hat{\mathbf{S}}\hat{\mathbf{D}}^{-\frac{1}{2}}\mathbf{H}^{(1)}\mathbf{W}^{(1)}\right),\end{aligned} \tag{5}$$

where $\hat{\mathbf{S}} = \mathbf{S} + \mathbf{I}_n$ and $\hat{\mathbf{D}}_{ii} = \sum_j \hat{\mathbf{S}}_{ij}$. For semi-supervised multilabel classification, we evaluate the mean square error over all labeled samples:

$$\text{Mean Square Error} = \frac{1}{t}\sum_{i=1}^{t}\left(\mathbf{H}_i^{(2)} - \mathbf{Y}_i\right)^2, \tag{6}$$

where $\mathbf{Y} \in [0,1]^{n \times c}$ is the ground truth label matrix with c labelsets, and t is the number of labeled samples.

4. Experiments

4.1 Datasets

The multilabel datasets used in this paper and their associated statistics are shown in **Table 1**.

4.2 Experimental setup

In this study, we have added probabilistic classifier chains [36], CSMLC [37] and RethinkNet [38] as baselines for comparison. The experimental settings are as follows: First, multilabel datasets are preprocessed to [0,1] as inputs, 80% of the

Datasets	Domain	# of features	# of samples	# of training data	# of test data	# of classes
Emotions*	Audio	72	593	474	119	6
Water Quality*	Chemistry	16	1060	848	212	14
CIE Image*	Image	294	2000	1600	400	5
Natural Scenes*	Image	294	2407	1925	482	6
Yeast*	Biology	103	2417	1933	484	14
AR Face**	Image	1024	30303	24242	6061	6

**Multilabel datasets are available at http://mulan.sourceforge.net/datasets-mlc.html*
***AR Face dataset is available at http://www2.ece.ohio-state.edu/_aleix/ARdatabase.html*

Table 1.
Statistics of the multilabel datasets.

samples are used for model (both multilabel learning and proposed method) training, and the last 20% of the samples are used as test sets. We also add Gaussian noise ranging from 6% to 12% of each test sample to test the robustness of the model. The overall framework is shown in **Figure 2**.

For deep learning, we train all models for 200 epochs using Adam [39] with a learning rate of 0.01 and the mean square error as the loss function.

4.3 Evaluation metrics

In multilabel learning, the evaluation metrics must be more rigorous than traditional single-label learning because one sample may be associated with multiple labels. These evaluation metrics [15] are divided into three groups, as shown in **Figure 3**. The higher the values of the F1 score, precision, mean average precision and recall, the better the performance is. The lower the values of the Hamming loss, one-error, coverage and ranking loss, the better the performance is. We consider the Hamming loss, one-error and mean average precision as three major metrics.

4.4 Experimental results

All experiments use different combinations of training and test data to verify the trained model and average the results after repeating the training ten times. According to the observations in **Figures 4–6**, the following conclusions are reached:

- Regardless of whether the Gaussian noise is added to the data set, the classification results of the problem transformation methods (BR, CCs, CAL, LP and RKL) are almost worse than the adaptive algorithms (MLkNN, ML-SVM and ML-DT)

- Deep learning may not obtain the best performance.

- We found that our method was raised on average by 1.8% and 8% higher in Hamming loss and mean average precision, respectively. And also has excellent performance even if the dataset were contaminated by noise.

- Regardless of whether noise is added to the data, our method in one-error evaluation is not as good as other baselines.

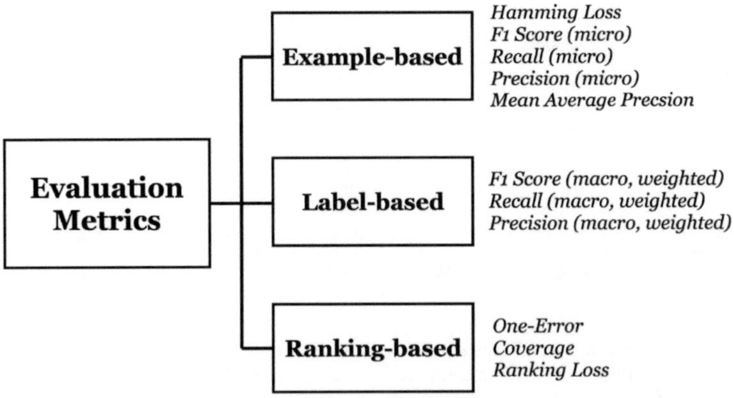

Figure 3.
Taxonomy of evaluation metrics.

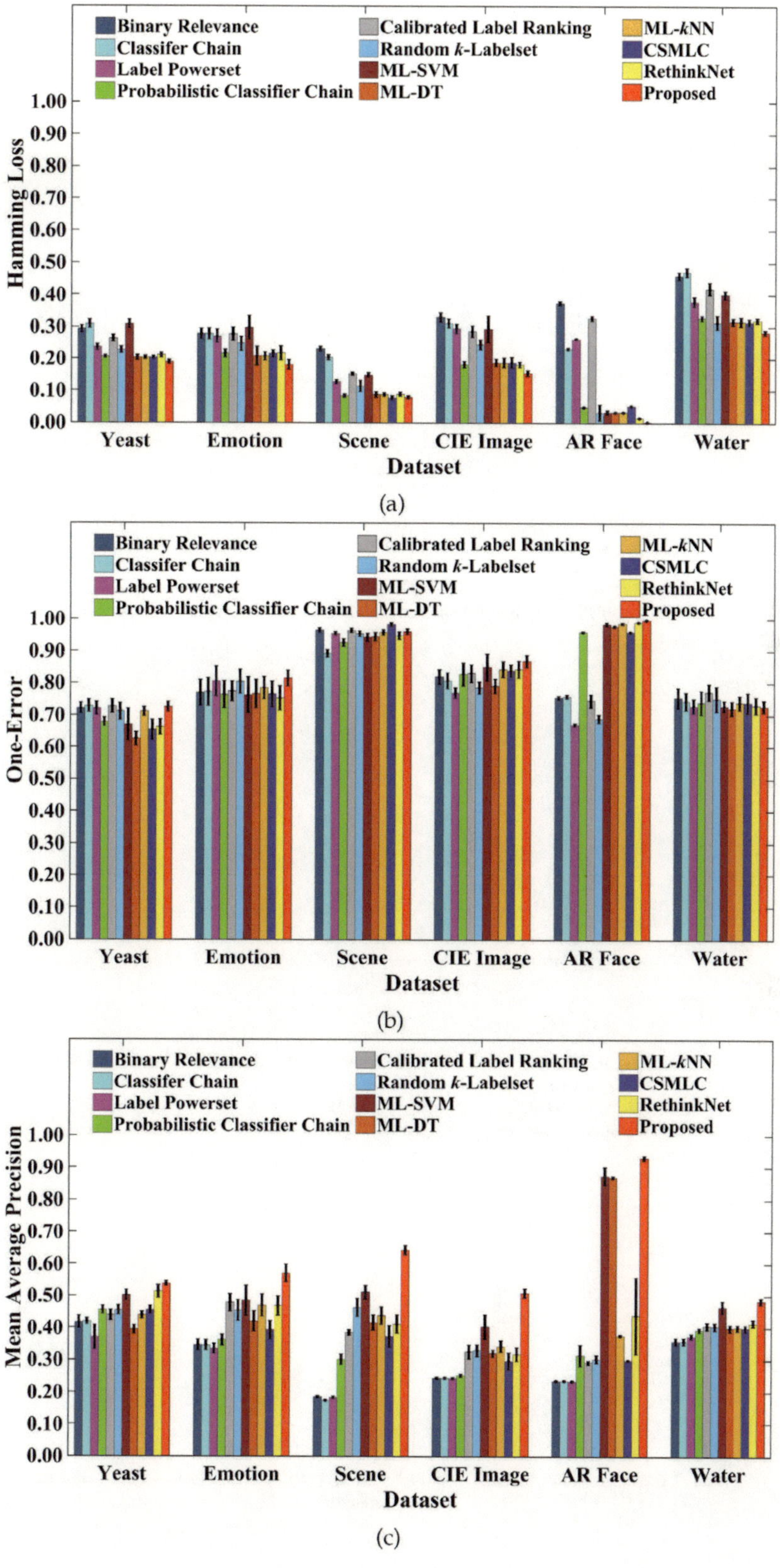

Figure 4.
Results of the proposed method compared with multilabel learning algorithms on the used multilabel datasets. (a)–(c) show the results without adding Gaussian noise.

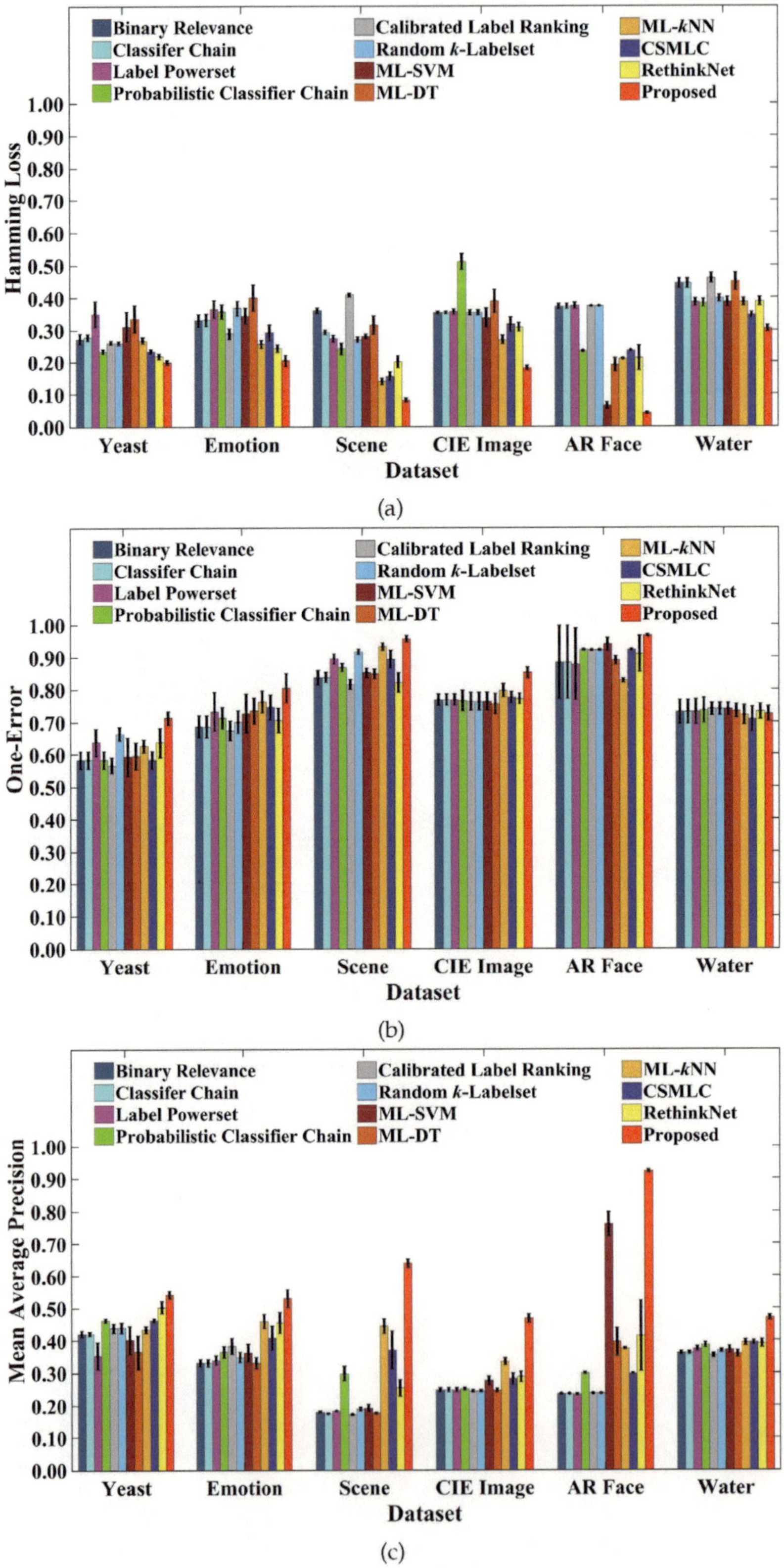

Figure 5.
Results of the proposed method compared with multilabel learning algorithms on the used multilabel datasets. (a)–(c) show the results of adding 6% Gaussian noise.

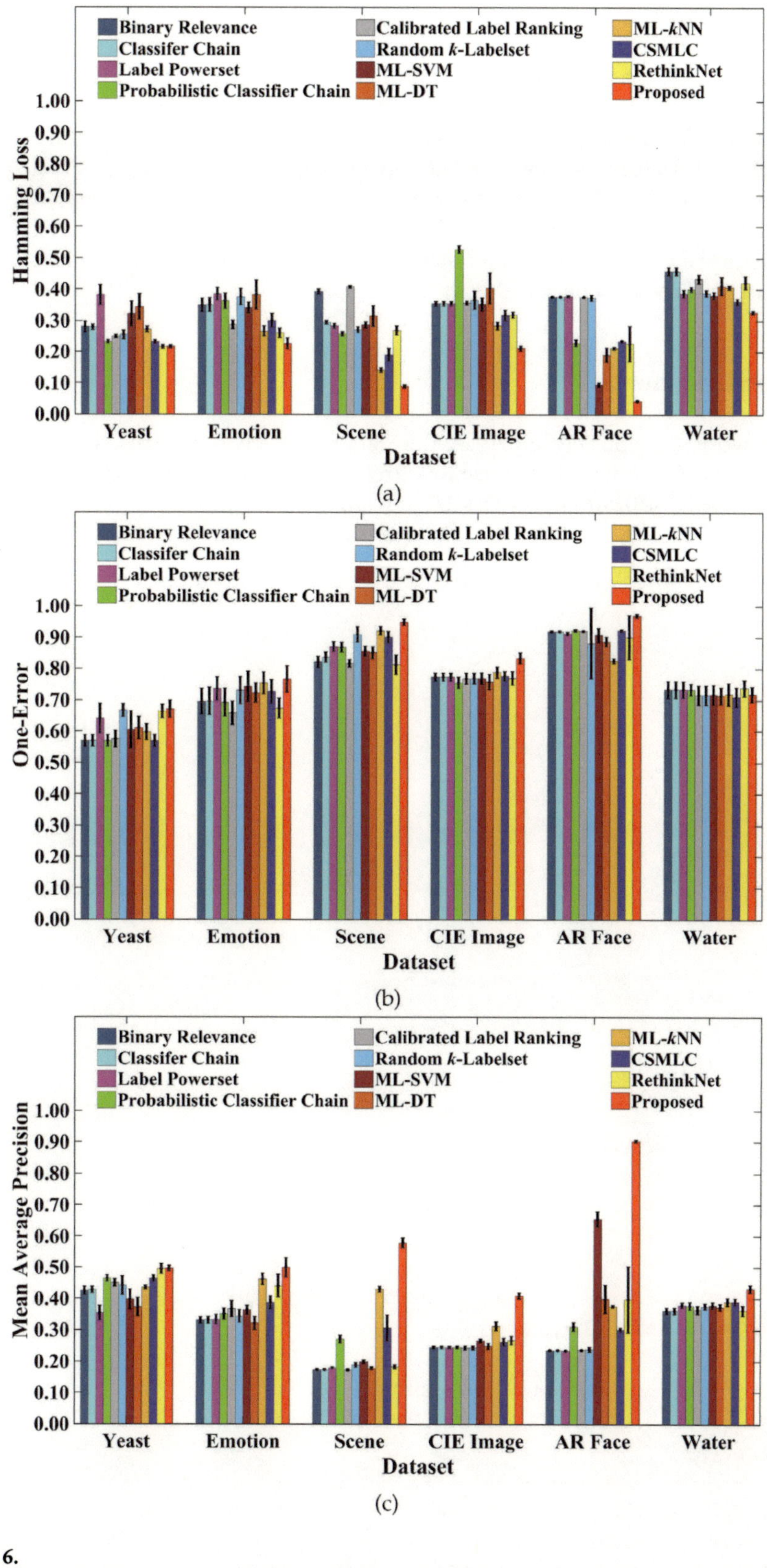

Figure 6.
Results of the proposed method compared with multilabel learning algorithms on the used multilabel datasets. (a)–(c) show the results of adding 12% Gaussian noise.

5. Conclusions

In this paper, we proposed a method of constructing a relation matrix by considering the correlation and sparsity of paired samples. We then added the characteristics of a GCN, which aggregates similar samples, to finally obtain the probability of occurrence of each label. Experimental results on six datasets showed that our proposed method can deliver superior performance in comparison with eleven baselines. Our future work will include designing a general framework that can reduce the use of memory and increase the efficiency of a GCN and extending this framework to unsupervised learning.

Acknowledgements

This work is supported in part by the Data Science Lab, NSYSU, and in part by the Pervasive Artificial Intelligence Research (PAIR) Lab, Taiwan, under the grant Nos. 110-2634-F-008-004 and 110-2221-E-110-046.

Author details

Wei-Cheng Ye[1] and Jia-Ching Wang[2*]

1 Department of Electrical Engineering, National Sun Yat-sen University, Kaohsiung, Taiwan

2 Department of Computer Science and Information Engineering, National Central University, Taoyuan, Taiwan

*Address all correspondence to: jcw@csie.ncu.edu.tw

References

[1] R. E. Schapire and Y. Singer, "BoosTexter: A boosting-based system for text categorization," *Machine Learning*, vol. 23, no. 3, pp. 135–168, May 2020.

[2] H. Peng, J. Li, S. Wang, L. Wang, Q. Gong, R. Yang, B. Li, P. S. Yu, and L. He,"Hierarchical taxonomy-aware and attentional graph capsule RCNNs for large-scale multi-label text classification," *IEEE Transactions on Knowledge and Data Engineering*, vol. 33, no. 6, pp. 2505–2519, Jun. 2021.

[3] M. R. Boutell, J. Luo, X. Shen, and C. M. Brown, "Learning multi-label scene classification," *Pattern Recognition*, vol. 37, no. 9, pp. 1757–1771, May 2004.

[4] L. Chen, W. Zhan, W. Tian, Y. He, and Q. Zou, "Deep integration: A multi-label architecture for road scene recognition," *IEEE Transactions on Image Processing*, vol. 28, no. 10, pp. 4883–4898, Oct. 2019.

[5] B.-W. Chen, M. Imran, N. Nasser, and M. Shoaib, "Self-aware autonomous city: From sensing to planning," *IEEE Communications Magazine*, vol. 57, no. 4, pp. 33–39, Apr. 2019.

[6] W. Ji, J. Xu, H. Qiao, M. Zhou, and B. Liang, "Visual IoT: Enabling internet of things visualization in smart cities," *IEEE Network*, vol. 33, no. 2, pp. 102–110, Mar.–Apr. 2019.

[7] W. Ji, B. Liang, Y. Wang, R. Qiu, and Z. Yang, "Crowd V-IoE: Visual internet of everything architecture in AI-driven fog computing," *IEEE Wireless Communications*, vol. 27, no. 2, pp. 51–57, Apr. 2020.

[8] B.-W. Chen, "Novel kernel orthogonal partial least squares for dominant sensor data extraction," *IEEE Access*, vol. 8, pp. 36131–36139, Feb. 2020.

[9] H. Chuang, K.-L. Hou, S. Rho, and B.-W. Chen, "Cooperative comodule discovery for swarm-intelligent drone arrays," *Computer Communications*, vol. 154, pp. 528–533, Mar. 2020.

[10] B.-W. Chen, "Symmetric nonnegative matrix factorization based on box-constrained half-quadratic optimization," *IEEE Access*, vol. 8, pp. 170976–170990, Sep. 2020.

[11] W. Ji, L.-Y. Duan, X. Huang, and Y. Chai, "Astute video transmission for geographically dispersed devices in visual IoT systems," *IEEE Transactions on Mobile Computing*, Jul. 2020.

[12] A. Elisseeff and J. Weston, "A kernel method for multi-labelled classification," in *Proc. International Conference on Neural Information Processing Systems: Natural and Synthetic*, Vancouver, British Columbia, Canada, 2001, Dec. 03–08, pp. 681–687.

[13] Y.-X. Li, S. Ji, S. Kumar, J. Ye, and Z.-H. Zhou, "Drosophila gene expression pattern annotation through multi-instance multi-label learning," *IEEE/ACM Transactions on Computational Biology and Bioinformatics*, vol. 9, no. 1, pp. 98–112, Feb. 2012.

[14] Z. Wu, S. Pan, F. Chen, G. Long, C. Zhang, and P. S. Yu, "A comprehensive survey on graph neural networks," *IEEE Transactions on Neural Networks and Learning Systems*, pp. 1–21, Mar. 2020.

[15] M.-L. Zhang and Z.-H. Zhou, "A review on multi-label learning algorithms," *IEEE Transactions on Knowledge and Data Engineering*, vol. 26, no. 8, pp. 1819–1837, Aug. 2014.

[16] F. Scarselli, M. Gori, A.-C. Tsoi, M. Hagenbuchner, and G. Monfardini, "The graph neural network model,"

IEEE Transactions on Neural Networks, vol. 20, no. 1, pp. 61–80, Jan. 2009.

[17] A. Krizhevsky, I. Sutskever, and G. E. Hinton, "ImageNet classification with deep convolutional neural networks," in *Proc. International Conference on Neural Information Processing Systems*, Lake Tahoe, Nevada, United States, 2012, Dec. 03–06, pp. 1097–1105.

[18] X. Mai, H. Zhang, X. Jia, and M. Q.-H. Meng, "Faster R-CNN with classifier fusion for automatic detection of small fruits," *IEEE Transactions on Automation Science and Engineering*, vol. 17, no. 3, pp. 1555–1569, Jul. 2020.

[19] R. Li, Q. Liu, J. Gui, D. Gu, and H. Hu, "Indoor relocalization in challenging environments with dual-stream convolutional neural networks," *IEEE Transactions on Automation Science and Engineering*, vol. 15, no. 2, pp. 651–662, Apr. 2018.

[20] P. Sen, G. Namata, M. Bilgic, L. Getoor, B. Galligher, and T. Eliassi-Rad, "Collective classification in network data," *AI Magazine*, vol. 29, no. 3, p. 93, Sep. 2008.

[21] F. Monti, D. Boscaini, J. Masci, E. Rodolà, J. Svoboda, and M. M. B., "Geometric deep learning on graphs and manifolds using mixture model CNNs," in *Proc. IEEE Conference on Computer Vision and Pattern Recognition*, Honolulu, Hawaii, United States, 2017, Jul. 21–26, pp. 5425–5434.

[22] W. Liu, S. Fu, Y. Zhou, Z.-J. Zha, and L. Nie, "Human activity recognition by manifold regularization based dynamic graph convolutional networks," *Neurocomputing*, vol. 444, pp. 217–225, Jul. 2021.

[23] Z.-M. Chen, X.-S. Wei, P. Wang, and Y. Guo, "Multi-label image recognition with graph convolutional networks," in *Proc. IEEE/CVF Conference on Computer Vision and Pattern Recognition*, Long Beach, California, United States, 2019, Jun. 15–20, pp. 5172–5181.

[24] X. Zhang, C. Xu, and D. Tao, "Context aware graph convolution for skeleton-based action recognition," in *Proc. IEEE/CVF Conference on Computer Vision and Pattern Recognition*, Seattle, United States, 2020, Jun. 14–19, pp. 14333–14342.

[25] R. Zeng, W. Huang, M. Tan, Y. Rong, P. Zhao, J. Huang, and C. Gan, "Graph convolutional networks for temporal action localization," in *Proc. IEEE/CVF International Conference on Computer Vision*, Seoul, Korea, 2019, Oct. 27–Nov. 02, pp. 7094–7103.

[26] L. Yao, C. Mao, and Y. Luo, "Graph convolutional networks for text classification," in *Proc. AAAI Conference on Artificial Intelligence*, Honolulu, Hawaii, United States, 2019, Jan. 27–Feb. 01, pp. 7370–7377.

[27] R. Ragesh, S. Sellamanickam, A. Iyer, R. Bairi, and V. Lingam, "Hetegcn: Heterogeneous graph convolutional networks for text classification," in *Proc. ACM International Conference on Web Search and Data Mining*, New York, United States, 2021, Mar. 08–12, pp. 860–868.

[28] X. Zhou, F. Shen, L. Liu, W. Liu, L. Nie, Y. Yang, and H. T. Shen, "Graph convolutional network hashing," *IEEE Transactions on Cybernetics*, vol. 50, no. 4, pp. 1460–1472, Apr. 2020.

[29] R. Xu, C. Li, J. Yan, C. Deng, and X. Liu, "Graph convolutional network hashing for cross-modal retrieval," in *Proc. International Joint Conference on Artificial Intelligence*, Macao, China, 2019, Aug. 10–16, pp. 982–988.

[30] D. Hong, L. Gao, J. Yao, B. Zhang, A. Plaza, and J. Chanussot, "Graph convolutional networks for hyperspectral image classification,"

IEEE Transactions on Geoscience and Remote Sensing, pp. 1–13, Aug. 2020.

[31] A. Qin, Z. Shang, J. Tian, Y. Wang, T. Zhang, and Y. Y. Tang, "Spectral–spatial graph convolutional networks for semisupervised hyperspectral image classification," *IEEE Geoscience and Remote Sensing Letters*, vol. 16, no. 2, pp. 241–245, Feb. 2019.

[32] H. Wang, Y. Yang, and B. Liu, "GMC: Graph-based multi-view clustering," *IEEE Transactions on Knowledge and Data Engineering*, vol. 32, no. 6, pp. 1116–1129, Jun. 2019.

[33] J. Wright, A. Y. Yang, A. Ganesh, S. S. Sastry, and Y. Ma, "Robust face recognition via sparse representation," *IEEE Transactions on Pattern Analysis and Machine Intelligence*, vol. 31, no. 2, pp. 210–227, Feb. 2009.

[34] T. N. Kipf and M. Welling, "Semi-supervised classification with graph convolutional networks.," in *Proc. International Conference on Learning Representations*, Toulon, France, 2017, Apr. 24–26.

[35] D. I. Shuman, S. K. Narang, P. Frossard, A. Ortega, and P. Vandergheynst, "The emerging field of signal processing on graphs: Extending high-dimensional data analysis to networks and other irregular domains," *IEEE Signal Processing Magazine*, vol. 30, no. 3, pp. 83–98, May 2013.

[36] K. Dembczyński, W. Cheng, and E. Hüllermeier, "Bayes optimal multilabel classification via probabilistic classifier chains," in *Proc. International Conference on Machine Learning*, Haifa, Israel, 2010, Jun. 21–24, pp. 279–286.

[37] K.-H. Huang and H.-T. Lin, "Cost-sensitive label embedding for multi-label classification," *Machine Learning*, vol. 106, no. 9–10, pp. 1725–1746, Oct. 2017.

[38] Y.-Y. Yang, Y.-A. Lin, H.-M. Chu, and H.-T. Lin, "Deep learning with a rethinking structure for multi-label classification," in *Proc. Asian Conference on Machine Learning*, Nagoya, Japan, 2019, Nov. 17–19, pp. 125–140.

[39] D. P. Kingma and J. Ba, "Adam: A method for stochastic optimization," in *Proc. International Conference for Learning Representations*, San Diego, California, United States, 2015, May 07–09.

Chapter 5

DMAPT: Study of Data Mining and Machine Learning Techniques in Advanced Persistent Threat Attribution and Detection

P.V. Sai Charan, P. Mohan Anand and Sandeep K. Shukla

Abstract

Modern-day malware is intelligent enough to hide its presence and perform stealthy operations in the background. Advance Persistent Threat (APT) is one such kind of malware attack on sensitive corporate and banking networks to stay there for a long time undetected. In real-time corporate networks, identifying the presence of intruders is a big challenging task for security experts. Recent APT attacks like Carbanak, The Big Bang, and Red Echo attack (targeting the Indian power sector) are ringing alarms globally. New data exfiltration methods and advancements in malware techniques are the two main reasons for rapid and robust APT evolution. Although many traditional and hybrid methods are available to detect this stealthy malware, the number of target-specific attacks are increasing rapidly at global level. Attackers have been crafting payloads resistant to malware sandbox environments so that traditional sandboxing techniques may not work with these APT malware detection. In this paper, we shed light on various Data Mining, Machine Learning techniques and frameworks used in both Attribution and Detection of APT malware. Added to this, our work highlight GAP analysis and need for paradigm shift in existing techniques to deal with evolving modern APT malware.

Keywords: APT, Targeted Malware, Data Exfiltration, APT Attribution

1. Introduction

Recent advances in the design of sophisticated malware tools are posing a significant challenge not only to the global IT industry but also to the banking and security organisations. Advanced Persistent Threat (APT) is a key player in highly targeted and sophisticated state-sponsored attacks [1]. These APT groups design and deploy malware in a unique way depending on the target. After selecting the targeted organisation, they come with different Tools, Techniques and Procedures (TTP) to bypass the traditional line of defences (intrusion detection systems or firewall). Once they get access, these APT groups stay inside targeted networks for a long time to observe the workflows. These APT groups use intelligent multi-stage malware deployment techniques to stay low under the radar for a long time [2]. Finally, gathered sensitive information is pushed in small chunks to its external control and command servers (C2C) using some clever exfiltration techniques.

The whole process of the APT life cycle is broadly divided into seven different phases as shown in **Figure 1** [3]. In the Reconnaissance phase, the attacker chooses the target network and studies the internal network structure and comes up with the necessary strategy, TTP, to bypass the initial layer of defence. Reconnaissance is followed by the Initial compromise phase, where attackers exploit open vulnerabilities to get an initial foothold into the targeted network. After that, the attackers try to replicate and propagate into another machine and establishes backdoors to pull more sophisticated payloads in Establishing foothold phase. Later in the Lateral movement phase, attackers escalate various privileges to perform more sophisticated tasks to hide its traces. In this particular phase, attackers traverse from one network to another network in search of sensitive information. After collecting the necessary data, the attackers strategically centralises this collected data to staging servers. In the data exfiltration phase, attackers use different custom encoding and encryption mechanisms to push these collected data to external control and command servers. Finally, to preserve the anonymity of the process, attacker leaves no traces by clearing the tracks and creates a backdoor to revisit that particular organisation in the future.

APT has grown to become a global tool for cyber warfare between countries. Carbanak APT campaign infected thousands of people worldwide and caused nearly $1 billion damage across the globe [4]. APT actors carried out a variety of actions in this operation, including opening fraudulent accounts and employing bogus services to obtain funds, as well as sending money to cybercriminals via the SWIFT (Society for Worldwide Interbank Financial Telecommunication) network. Similarly, in 2018, Big Bang APT developed a much more robust and sophisticated multi-stage malware targeting the Palestinian Authority [5]. This APT malware includes several modules that perform tasks ranging from obtaining a file list, capturing screenshots, rebooting the machine, retrieving system information, and self-deletion. More recently, a supply chain attack on solar winds by the Russian APT group was considered one of the sophisticated attacks. RefreshInternals() method in solar winds attack depict the maturity of these state-sponsored APT groups in terms of malware design and payload delivery [6].

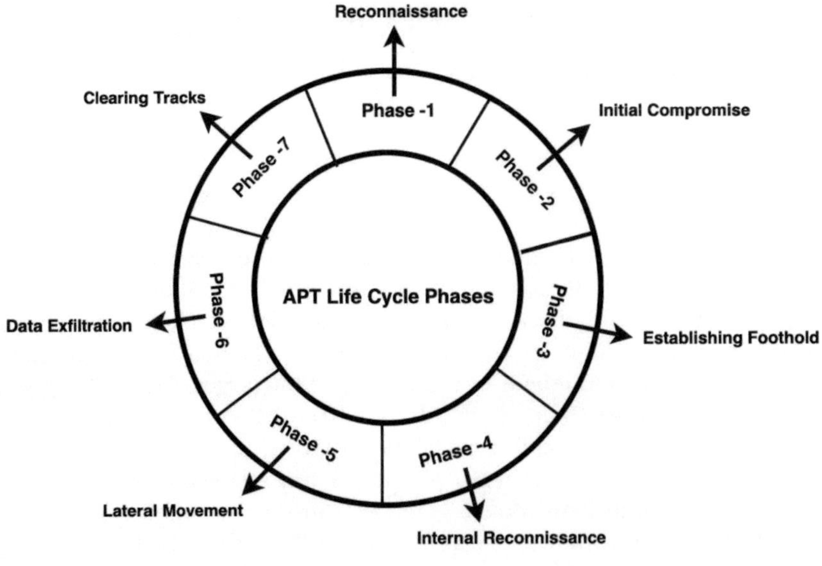

Figure 1.
APT life cycle phases.

In order to deal with these kinds of state-sponsored targeted attacks, security experts consider APT attribution and detection as two key pillars. Attribution is an analysis process that explains about "who" is behind particular cyber espionage and "why" they have done it [7]. This process gives insights about particular APT threat actors and their targeted areas as well. Based on this preliminary information, the security community try to detect these attacks by fixing issues at different levels of an organisation. Since APT attribution and detection became crucial for many security firms/govt agencies, both these processes require massive data pre-processing and analysis. To address these issues, researchers propose different data mining and machine learning techniques in both attribution and detection as well. In this paper, we discuss various data mining and machine learning techniques in both detection and attribution of APT malware. In addition to this, we compare different detection techniques, and we highlight research gaps among those techniques which need to be addressed by the security community to combat this sophisticated APT malware.

This paper is organised as follows. Section 1. details APT overview and phases of APT, followed by the need for data mining and ML techniques in both attribution and detection of APT malware. Section 2. talks about the process of attribution and different techniques proposed to perform APT attribution. Section 3. discuss about various state of the art data mining and ML techniques proposed by the research community in APT detection. Section 4. details research gap analysis followed by conclusion and future scope.

2. Data mining and ML techniques in APT attribution

APT attribution is an analysis process that reveals the identity of the threat actors and their motto through a series of steps [8]. First, security firms collect data from different victim organisations by performing forensic analysis on the respective networks and collect different Indicators of Compromise (IOC). In general, attackers repeat this pattern in several other organisations as well. Security firms observe and analyse these repeated patterns in IOC and TTP's together, and cluster these combinations as intrusion sets. Performing data analytics on these intrusion sets over a period will eventually reveal the threat actor and motivation behind the attack as depicted in **Figure 2**, respectively.

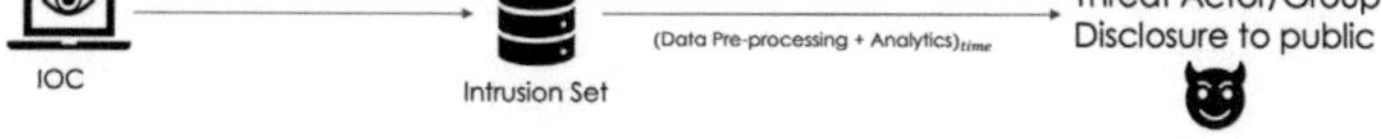

Figure 2.
Overview of APT attribution process.

2.1 DeepAPT: APT attribution using deep neural network and transfer learning

APT attribution is quite a challenging task to the security community because of various reasons. Majorly, State-sponsored APTs are developed in the supervision of different units and equipped with default Anti-VM and Anti-Debugging techniques to obfuscate the payloads. This technique makes feature extraction extremely challenging to most security firms. In addition to this, APT malware samples are highly targeted so that very few samples will be available for analysis purposes. In order to address this issue, Rosenberg et al. proposed a technique for APT attribution by using a Deep Neural Network (DNN) classifier [9]. In this research work, the authors used 3200 malware samples for training DNN classifiers, 400 samples for validation and 1000 samples for testing the model. All the APT malware samples are executed in a cuckoo sandbox environment, and generated reports are used as raw input in training the classifier. DNN is effective in learning high-level features on its own from raw inputs. In order to train DNN models more effectively, in this work, the authors removed top 50,000 frequent words from input features of all cuckoo reports. So, DNN models take very uncommon words from all cuckoo reports and build a much more effective model to perform APT attribution. This DNN architecture is a 10-layer, fully connected network (50,0000 neurons at the input layer and 2,000 in the first hidden layer) with a ReLU activation function. The final trained APT attribution model did decent work on test data with 98.6% accuracy. Added to this, the authors also applied transfer learning on trained DNN models by removing and retraining top layer neurons. After applying transfer learning, the model still performs exceptionally well with 97.8% accuracy. From the t-distributed stochastic neighbour embedding algorithm (used to reduce from 500 dimension space to 2 dimension space), we can see that the trained model could separate different APT malware groups as shown in **Figure 3**, respectively.

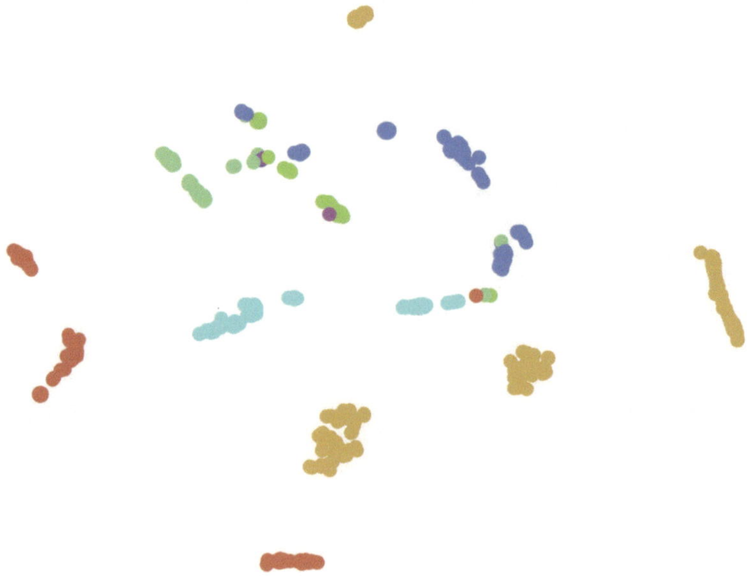

Figure 3.
2-dimensional visualisation of APT families using t-SNE algorithm [9].

2.2 APT attribution based on threat intelligence reports

Most APT attribution techniques heavily rely on performing analysis for malware samples used in that particular campaign. The key disadvantage of this strategy is that the same malware samples can be used in several operations. In some situations, APT groups specifically buy malware from the dark web based on their requirements. So, the ML models constructed by only considering malware samples may not give efficient results in terms of APT attribution. In order to address this issue, Lior Perry et al. proposed a method named NO-DOUBT, i.e. Novel Document Based Attribution, by constructing models on threat intelligence reports with the help of Natural Language Processing (NLP) techniques [10]. In this research, the authors collected 249 threat intelligence reports of 12 different APT actors and considered APT attack attribution as a multi-text classification problem. The proposed model consists of mainly two phases, as shown in **Figure 4**. In the training phase, labelled reports and word embeddings transform the input data to a vector representation. For generating this vector representation, authors propose SMOBI (Smoothed Binary Vector) algorithm, which will find cosine similarities between input words in labelled data sets and word embeddings to form a huge n × m matrix. This vector representation and labels are given to the ensemble xGBoost classifier to construct a known actor model. In the deployment phase, new test reports (unlabelled) are also converted to vector representation and given to the known actor model to determine the probability predictions to the known classes. These probability predictions are given to a New Actor Model (a binary classifier that outputs whether it is a known APT actor or a new unknown actor) to make final predictions. Although this model struggles to detect Deep Panda and APT29 actors, SMOBI based APT attribution outperforms previous text-based APT attribution models (unigrams + bigrams and tf-idf) in terms of Accuracy, Precision and Recall.

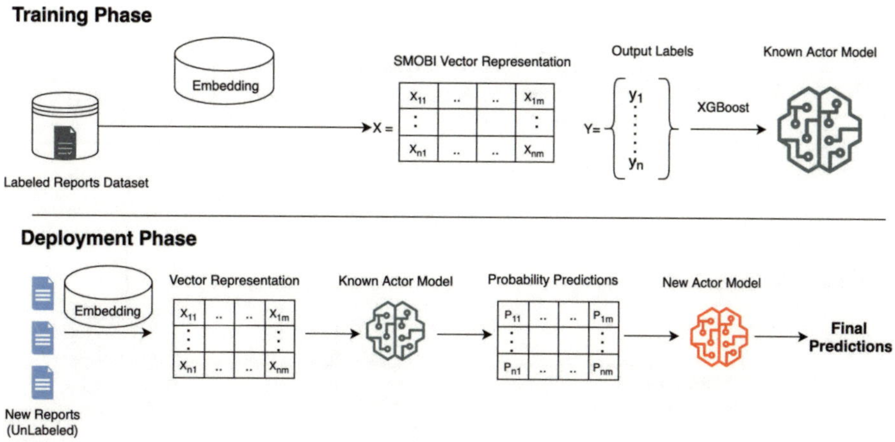

Figure 4.
NO-DOUBT method for APT attribution [10].

2.3 ML based attribution framework using high level IOC

Most of the APT attribution processes depend upon the manual analysis in victim networks and collecting low-level indicators of compromise (forensic analysis at firewalls, tracebacks, IDS and Honeypots). However, APT actors change this low-level IOC from one organisation to another organisation. ML models built based on this low-level IOC, results in inadequate cyber intelligence systems. On the other hand, collecting high-level IOC's for each organisation is time-consuming. Such high-level IOC's are published in the form of Cyber Threat Intelligence (CTI) reports across the organisations as a common practice. In 2019, Umara Noor et al. proposed a distributional semantic technique of NLP to build a cyber threat attribution framework by extracting patterns from CTI reports [11]. The proposed attribution framework is broadly divided into three phases, as depicted in **Figure 5**. In this experiment, authors used a customised search engine to collect 327 unstructured CTI documents corresponding to 36 APT actors as a part of data collection phase. The CTI documents do not contain the exact keyword described in the standard taxonomy due to varying textual definitions and choices for communicating a concept. Rather than using a

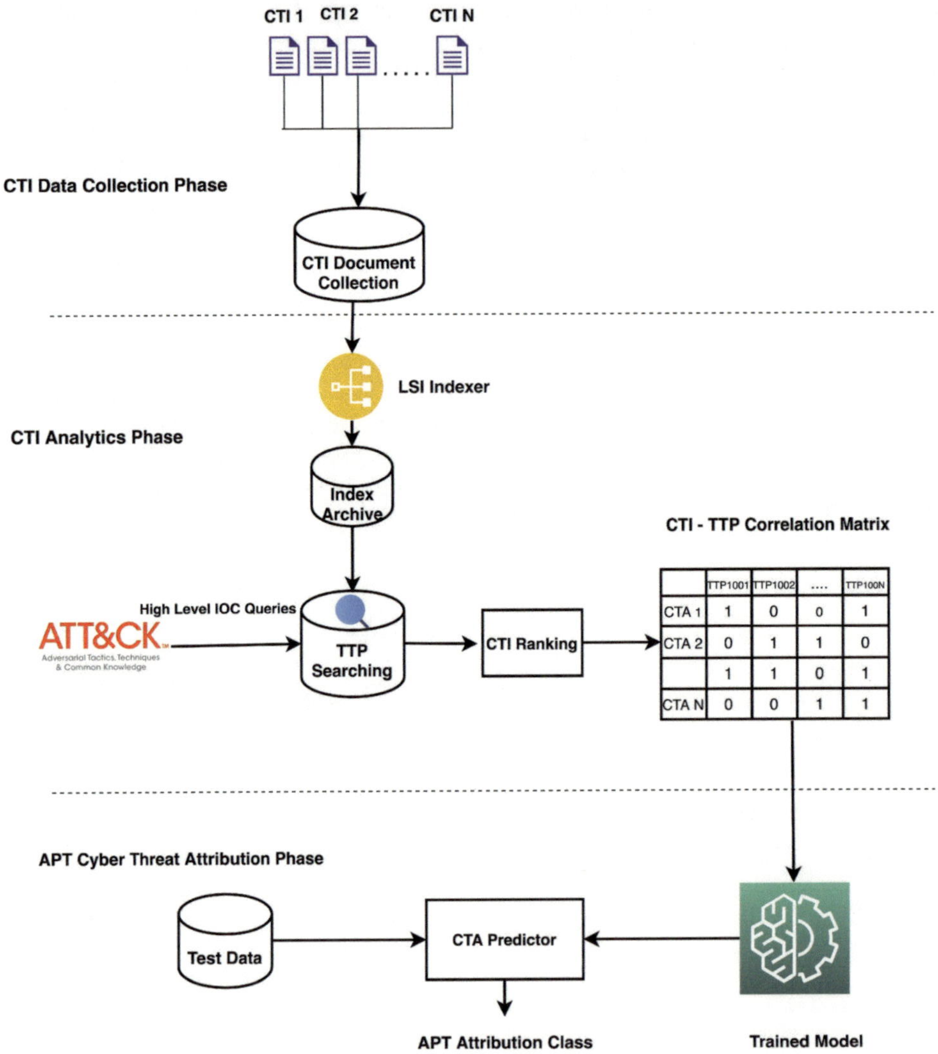

Figure 5.
Cyber threat attribution framework [11].

simple keyword-based search, the authors developed a semantic search method based on the statistical distributional semantic relevance technique (LSA), to retrieve relevant documents. The input CTI records are indexed using LSA. The statistically derived conceptual indices (from LSA indexer) are searched for semantically relevant topics using the high-level IOC labels specified in MITRE ATT&CK [11]. Based on cosine similarity, the CTA-TTP correlation matrix is constructed in the CTI analytics phase. ML models are built on top of the CTA-TTP correlation matrix in the Cyber Threat Attribution phase. Among various classifiers, the Deep Neural Network turned out to be the best performer with 94% attribution accuracy on test data with high precision and recall values.

2.4 APTMalInsight: recognising APT malware based on system call information and ontology framework

Behavioural analysis of APT malware gives better insights on both APT attribution and detection. Based on this motivation, Weijie Han et al. proposed that, dynamic system call information reveals behavioural characteristics of APT malware [12]. Furthermore, the authors built an ontology model to understand in-depth relation between the maliciousness of APT malware to its families, as depicted in **Figure 6**, respectively. APTMalInsight framework mainly consists of two modules i.e. APT malware family classification module and detection module. The basic concept behind the APTMalInsight framework is to profile the behavioural characteristics of APT malware. It obtains dynamic system call information from the programs to reliably detect and attribute APT malware to their respective families. Primarily, APT malware samples are executed to extract dynamic API calls. After extracting API calls, authors calculated the feature importance of each API call and built a feature vector by selecting top N-API calls from the API call sequence. ML models built on top of that feature vector will output the APT attribution class for test data, as shown in **Figure 7**. For the experiment, authors considered a total of 864 APT malware samples belonging to five different families. As per the experimentation results, Random Forest turned out to be the best model in terms of Accuracy(98%), Precision and Recall for APT malware family attribution.

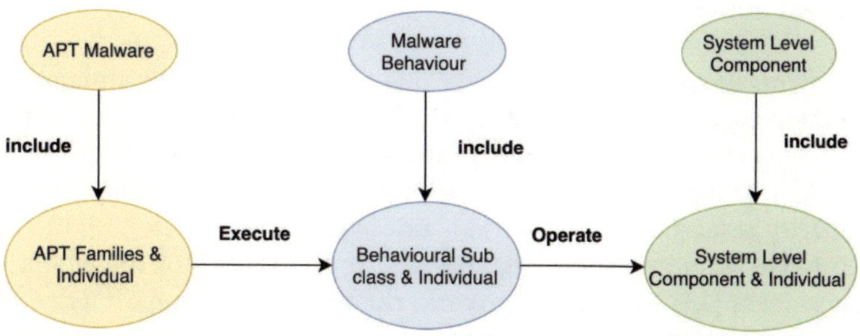

Figure 6.
APT malware ontology model [12].

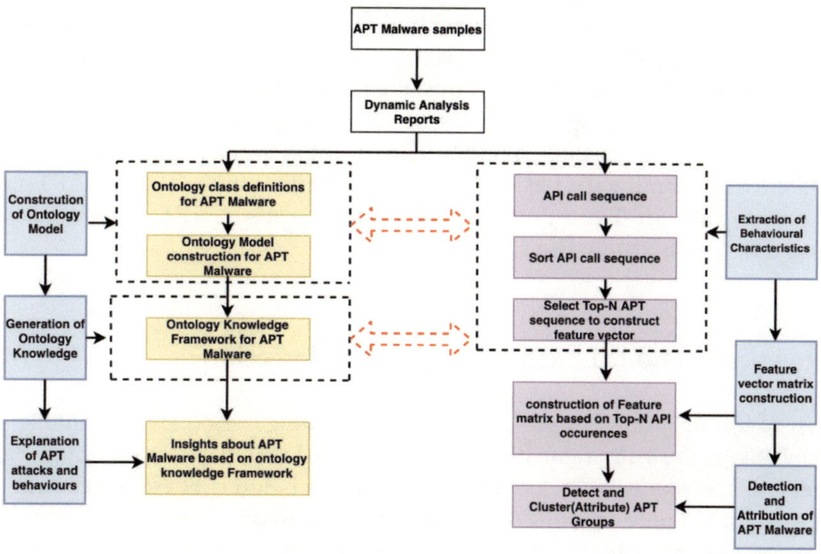

Figure 7.
High-level overview of APTMalInsight framework [12].

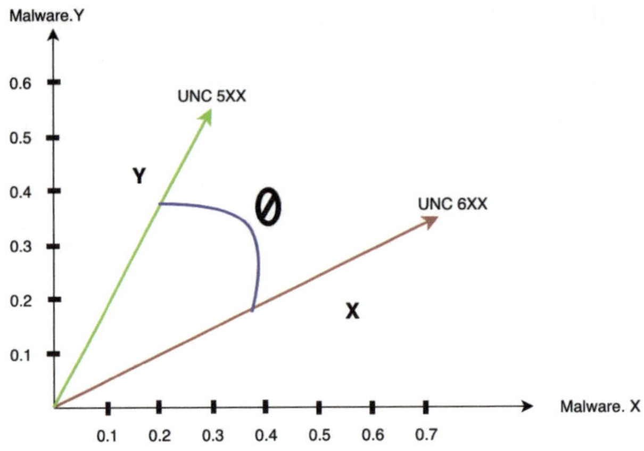

Figure 8.
Cosine similarity between different un-attributed APT groups [13].

2.5 ATOMIC: FireEye's framework for large scale clustering and associating APT threat actors

Security firms like FireEye investigate many victim networks and collect IOC and group them together as uncategorised ("UNC") intrusion sets. Over time, this type of UNC sets are increasing rapidly, and security firms need to either merge these other APT groups or assign a new group name based on manual analysis. FireEye security researchers proposed an automated framework with the help of ML models to perform investigation, analysis, and rationale for the whole APT attribution process [13]. In this framework, the researchers suggest a document clustering approach using term frequency and - inverse document frequency method (TF-IDF). The TF-IDF algorithm assigns more importance to a term if the word often appears in the document. Similarly, if the term appears common across all the documents, the algorithm decreases its importance. This method favours unique terms like custom malware families, which may appear in just a few classes, and downplays popular terms like 'phishing', which appear more often. After calculating scores using the TF-IDF algorithm, each UNC group is converted into a vector representation, and researchers calculate cosine similarity between these APT groups as shown in **Figure 8**, respectively. As angle between the two vectors decreases, they tend to become parallel. The decrease in the angle helps the researchers to determine the extent of similarity between two different APT groups. Based on this idea, FireEye automated the whole process of APT attribution and merging different uncategorised groups.

3. Data mining and ML techniques in APT detection

Most of the APT families stay undetected for a long period and use intelligent ways to damage the vulnerable hosts. When a traditional malware executes, most of the events occur sequentially and leave some traces behind. These traces help modern-day intelligent systems like SIEM, IDS, IPS to prevent these attacks. But, when it comes to the case of APTs, they clean the attack traces and also prevent sequence execution of events. Also, APT employs Anti-VM and Anti-debugging techniques for making things harder for the detection systems. The hardness in detecting the APT has made the cyber security enthusiasts draw their attention towards this domain. Some of the important contributions in the research area are mentioned below. A detailed comparison among different detection techniques are illustrated in **Table 1**, respectively.

3.1 A novel deep learning stack for APT detection

Tero et al. [14] proposed a theoretical approach for detecting APT by developing a stack of Deep Learning methods where each layer has a particular task in handling APT events. The authors consider network payload and packet header information as features, and they streamlined the input to the detection stack without any data filtering mechanism. The detection stack is designed sequentially. The initial layers, i.e. layer-1 and layer-2, are used to detect the known attacks and legitimate network traffic from the data flow respectively. Layer-3 of the detection stack employs in identifying the outliers having historical presence. It uses Recurrent Neural Network-Long Short Term Memory (RNN-LSTM) units to confirm whether an outlier has historical occurrence. Layer-4 helps to classify the outliers into four categories, i.e. regular traffic, known attack, predicted attack and unknown outlier

Research item	DM/ML technique employed in APT detection	Input data	Novelty
[14]	RNN-LSTM and GHSOM	Network Traffic Flow	Deep learning stack with sequential neural networks to detect APT.
[15]	Provenance Graph Mining	Host Audit Data (Linux audit or Windows ETW)	Suspicious information flows are identified using MITRE ATT&CK framework.
[16]	Directed Graph Mining and One Class SVM	SIEM Event Logs	Extracting attack vectors from SIEM logs
[17]	Continuous Association Rule Mining Algorithm (CRAMA)	IDS Logs	Identify correlation rules between various system events to develop an APT attack graph
[18]	RNN-LSTM	SIEM Event Logs	Identify possible event codes and their sequence to detect an APT attack in realtime
[19]	Ensemble Classifier	Network Traffic Flow	Separate threat detection sub-module for APT life cycle phases.
[20]	Multi fractal based error minimization	Network Traffic Flow	Multi fractal analysis to extract the hidden information of TCP connections.
[21]	Correlation Analysis	Multiple data sources	Construction of Attack Pyramid using multiple planes to detect APT
[22]	J48 Classifier	API log data	API calls to track process injection and privilege escalation activities.
[23]	Ensemble Classifier	Domain Names (Alexa and data. netlab.360)	Identify malicious C2C communication using lexical features of domain names
[24]	Ensemble Classifier	Domain Names (Alexa and DGArchive)	Identify malicious C2C communication using lexical, network features of domain names

Table 1.
Comparison of different APT detection methods.

using an anomaly detection method named Growing Hierarchical Self-Organising Map (GHSOM). The stack's final layer helps to map the anomalies (i.e. interconnections between the outlier events) using a Graph Database (GDB). The proposed stack model is highly modular and was designed to perform dynamic detection of APT events with a decent detection accuracy. However, this detection system is complex in design and result in higher time complexity when dealing with massive data inputs.

3.2 Real-time APT detection through correlation of suspicious information flows (HOLMES)

HOLMES model of APT detection is strongly based on the principles of the APT kill chain model. The cyber kill chain model gives a higher-level overview of the sequence of events in successful APT espionage, i.e. reconnaissance, command and control communication, privilege escalation, lateral movement, data exfiltration, and trace removal. Audit data from various operating systems are converted to a common data representation format and passed as input to the proposed model in the initial step. Lower-level information flows are extracted from the audit data such as process, files, memory objects and network information etc. The core part of the proposed model is to map the lower-level information data flows to the phases of the APT-kill chain by constructing an intermediate layer. The intermediate layer is responsible for identifying various TTP's (Tools, Techniques, Procedures) from the low-level information data flow that correlates with respective phase of the APT life cycle. The authors considered around 200 TTP patterns based on MITRE ATT&CK framework [15]. The TTP patterns and noise filtering mechanism are employed in constructing a High-Level Scenario Graph (HSG) from which we can detect the APT attack with decent accuracy.

3.3 Anomaly detection in log data using graph databases and machine learning to defend advanced persistent threats

Schindler et al. proposed an APT detection engine based on the principles of APT kill chain phases [16]. In this work, SIEM logs were considered as data source. The correlation is identified between the event logs and the phases of APT kill chain. An adapted kill chain model is constructed to identify the possible attack vectors from the SIEM event logs. This model is implemented at two different levels.

Level-1 deals with graph-based forensic analysis where logs from different programs are aggregated based on timestamp to identify events with in the network. A directed graph is constructed from the multiple layers of event sequences. Each event sequence reveals whether the event flow matches with the partial/full phases of the APT kill chain.

Level-2 helps in identifying various anomalous activities using the Machine Learning approach. An ML classifier is constructed to make the model robust in detecting APT events along with the graph model. Authors considered "one-class SVM" as the classifier model and used windows logs, firewall logs, file audit logs of benign system programs as its data source. This model is expected to identify all the events that differ from the benign programs.

The proposed model achieved a decent accuracy score of 95.33% in detecting APT events. However, considering the case of smart malware where malicious programs mimic normal user behaviour, the proposed model tends to produce a relatively high false-positives.

3.4 A study on cyber threat prediction based on intrusion event for APT attack detection

Yong-Ho Kim et al. [17] proposed a theoretical model for APT detection that consider intrusion detection system logs as data source. From the IDS logs, correlation rules between various system events are identified to build an attack graph. Identifying the correlation between the intrusion detection logs helps in predicting the future attacks. In the initial phase, intrusion detection logs are collected and corresponding intrusion events were extracted. The extracted events are passed to different function blocks, each corresponding to a particular detection activity. One of the functional block identifies the single-directional i.e. (host to C2C interaction) and bi-directional (host to C2C, C2C to host) communication activities. Another block identifies the repetitive intrusion events and combines them as a single event to optimise the time and resource constraints. A correlation analysis block identifies the context of intrusion detection events and creates sequential rules based on the principles of 5 W and 1H (When, Where, Why, Who, What and How). Finally, the prediction engine consider the attack scenario and tries to predict one or more events that can occur after a single intrusion event. This module consider data mining principles such as support and confidence to produce the best possible result. The time constraint is one of the practical problems with this model, as some of the functional blocks take a longer time to process events. Another important aspect is that, rules of the intrusion detection systems will directly affect the outcome of this model.

3.5 APT detection using long short term memory neural networks

Charan et al. [18] proposed an APT detection engine that takes SIEM event logs as input and use LSTM neural networks to detect the successful APT espionage. The author consider Splunk SIEM logs as a data source and streamline data to the Hadoop framework to process and obtain the event codes for every activity. Based on the APT life cycle phases, the author listed out the possible event codes and their sequence, leading to successful APT espionage. The core part of this work is to identify the event codes occurring in a sequence, and this process requires memorising the previous state event codes. So, in the proposed model, LSTM (a variant of RNN) is considered a classifier because it overcomes vanishing gradient problem by remembering the previous state event codes to confirm APT attack presence. However, this model may suffer from a high false-positive rate when smart malware techniques are employed in crafting the APT attack.

3.6 MLAPT: detection of APT attacks using machine learning and correlation analysis

APT detection research mainly rely on the analysis of malware payload used in different phases of APT attack. This kind of approach result in high false positives in case of multi-stage malware deployment. In order to address this issue, Ghafir et al. proposed a model to detect multi-stage APT malware by using machine learning and correlation analysis (MLAPT) [19]. The MLAPT system is broadly divided into three modules, i.e. 1) Threat detection module, 2) Alert correlation module and 3) Prediction module. Initially, network traffic is passed to the Threat detection module in which authors built several submodules to detect multi-stage attacks. The Output alerts from the Threat detection module are passed to the Alert correlation module. Alert correlation module filters redundant alerts and clusters these alerts based on correlation time interval. The correlation indexing sub-module determines a given scenario is either a full APT scenario or sub-APT scenario based on alert correlation score. The prediction module consider sub APT scenarios and predict its probability of becoming a full APT scenario. Based on that prediction module, alerts are escalated to the network security team to stop this APT kill chain. The novelty of this research lies in the detection of APT across all life cycle phases. Added to this, the MLAPT system monitors and detects real-time APT attacks with a decent 81% Accuracy.

3.7 Detection of APT attacks using fractal dimensions

Detecting APT network patterns is a complex task as it tries to mimic the behaviour of regular TCP traffic. APT malware opens and closes TCP connections to its C2C servers like any other regular legitimate connection with a minimal data transfer to stay low under the radar. Single scale analysis does not extract the complexities of this kind of APT traffic and lowers the detection accuracy. Researchers found that current supervised ML models use euclidean based error minimization, which results in high false positives while detecting complex APT traffic. To address these issues, Sana Siddiqui et al. proposed an APT detection model using multi-fractal based analysis to extract the hidden information of TCP connections [20]. Initially, the authors considered 30% of labelled datasets and computed prior correlation fractal dimension values for normal and APT data points. Both these computed values are loaded into the memory before processing the remaining 70% unlabelled dataset. Each point in the remaining 70% dataset is added to both normal and APT labelled dataset, and posterior fractal dimension values are calculated in the next step. The absolute difference between prior and posterior values for both regular and APT samples are calculated to determine the closest cluster to the data point. If fd_anom (absolute difference between prior and posterior for APT sample) $\leq$ fd_norm (absolute difference between prior and posterior for normal sample), then that data point is classified as an APT sample and vice versa. As per the experimental observations, fractal dimension based ML models performs better in terms of accuracy (94.42%) than the euclidean based ML models.

3.8 APT detection using context-based detection framework

Paul Guira et al. proposed a conceptual framework known as the Attack Pyramid for APT detection [21]. In this approach, the goal of the attack (data exfiltration in most of the cases) should be identified and placed on top of the pyramid. Further more, the model identifies various planes such as user plane, application plane, network plane and physical plane where the possibility of attacks are maximised. From the proposed approach, one can identify the correlation between various events across different planes. In general, an APT attack span multiple planes as the attack life cycle progresses. So, it is possible to identify the attack contexts that span through multiple attack planes. Events from different sources, i.e. VPN logs, firewall logs, IDS logs, authentication logs, system event logs are passed as data source to the detection engine. From these logs, the context of attack is identified using correlation rules. In the next step, the suspicious activities are identified by matching the attack contexts using a signature database. This model requires updating signatures at regular intervals to identify new attack contexts in real-time scenarios.

3.9 APT detection system based on API log data mining

Chun-I Fan et al. [22] proposed a generalised way for APT detection using system calls log data. The model was built based on the principles of dynamic malware analysis where API call (system call) events were passed through a detection engine. The novelty of this work lies in the approach of handling the API calls. Modern APT malware is often used to create child processes or inject code into a new process to evade detection. Authors have created a program named "TraceHook" that monitors all the code injection activities. Tracehook outputs the API count for the executable samples (benign/malware), and a machine learning classifier model is constructed on top of the obtained API count values. The proposed model considers only six important DLLs to monitor and can be combined with other APT detection models to build a robust APT detection engine.

3.10 Ensemble models for C2C communication detection

Identifying and stopping a particular life cycle event can break the full APT cycle and minimise damage to a considerable proportion. Based on this idea, researchers proposed various methods to stop malicious C2C communication. Modern-day malware employed a new way to communicate with their C2C server with the help of Domain Generation Algorithms (DGA). DGA creates a dynamic list of domain names in which a few domain names are active for a limited amount of time. So, the malware communicates to a different C2C domain name for every successful communication. This practice helps the smart malware to avoid detection from the traditional antivirus, firewalls, and other network scanning software. Anand et al. [23] proposed a classification technique to detect character-based DGA, i.e. domain names are constructed by concatenating characters in a pseudo-random manner, for example, wqzdsqtuxsbht.com. In this method, author extracted various lexical-based features such as n-grams, character frequencies, and statistical features to build an ensemble classifier. The proposed model can detect character-based DGA domain names with a decent accuracy score of 97%. Charan et al. [24] proposed a similar technique to detect word-based DGA domain names where domain names are constructed by concatenating two or three words from dictionaries, for example crossmentioncare.com. In their model, the author consider lexical, statistical, network-based features to build an ensemble classifier. A combination of the above two models can detect the C2C communication activity with a decent accuracy.

4. Conclusion and future scope

Although the security community propose different techniques to detect APT malware, there is a clear gap between current detection mechanisms and APT groups evolution. APT attack detection is extremely difficult due to an unavailability of benchmark datasets for training and evaluation. Added to this, constant change in TTP usage by APT groups result in high false positives in terms of detection. Due to the persistent nature of APT campaigns, it is cumbersome to capture the data over a long period of time. This raises the issue of storing and processing such large amounts of data so that real time detection is still a challenging task to the security community. Many state of the art APT detection models can be bypassed using modern Load Off Land Binaries (LolBins) and process injection through fileless malware. Lately, targeted APT malware evolved into a new variant named smart malware which is highly modular, robust and intelligent enough to evade detection from state of the art ML techniques. Along with these issues, adversarial machine learning is a potential threat to the existing detection mechanisms. Some of the APT groups also started using GAN to modify the payloads in such a way to evade detection and attribution as well. In order to address these serious security concerns in APT detection and attribution, there is a need for benchmark datasets and robust ML models working at different levels of the APT kill chain.

Conflict of interest

The authors declare no conflict of interest.

Abbreviations

APT	Advanced Persistent Threat
C2C	Control and Command Server
LoLBin	Living Off Land Binaries
TTP	Tools, Techniques and Procedures
GAN	Generative Adversarial Networks
DGA	Domain Generation Algorithms
CTI	Cyber Threat Intelligence
IOC	Indicators of Compromise
SMOBI	Smoothed Binary vector
GHSOM	Growing Hierarchical Self Organising Map
RNN	Recurrent Neural Network
LSTM	Long Short Term Memory Neural Networks
SIEM	Security Information and Event Management
IDS	Intrusion Detection System
IPS	Intrusion Prevention System
ETW	Event tracing for windows
CTA	Cyber Threat Attribution
DLL	Dynamic Link Library

Author details

P.V. Sai Charan*†, P. Mohan Anand† and Sandeep K. Shukla†
Department of Computer Science and Engineering, Indian Institute of Technology, Kanpur, India

*Address all correspondence to: pvcharan@cse.iitk.ac.in

† These authors contributed equally.

References

[1] Stojanović, Branka, Katharina Hofer-Schmitz, and Ulrike Kleb." APT datasets and attack modeling for automated detection methods: A review." Computers & Security 92 (2020): 101734. DOI: https://doi.org/10.1016/j.cose.2020.101734

[2] Zhou, Peng, et al." Detecting multi-stage attacks using sequence-to-sequence model." Computers & Security 105 (2021): 102203. DOI: https://doi.org/10.1016/j.cose.2021.102203

[3] APT Security: What Are Advanced Persistent Threats?. [Internet]. 2020. Available from : https://securitytrails.com/blog /advanced-persistent-threats-apt [Accessed: 25 May 2021]

[4] Kaspersky Lab: The Great Bank Robbery: The Carbanak APT (Detailed Investigation Report). [Internet]. 2015. Available from : https://securelist.com/the-great-bank-robbery-the-carbanak-apt/6873/ [Accessed: 25 May 2021]

[5] The Big Bang APT Report. [Internet]. 2018. Available from: https://research.checkpoint.com/apt-attack-middle-east-big-bang/ [Accessed: 25 May 2021]

[6] Microsoft Internal Solorigate Investigation Update. [Internet]. 2020. Available from : https://blogs.microsoft.com/on-the-issues/2020/12/17/cyberattacks-cybersecurity-solarwinds-fireeye/ [Accessed: 25 May 2021]

[7] Steffens, Timo. Attribution of Advanced Persistent Threats. Springer Berlin Heidelberg, 2020. DOI : https://doi.org/10.1007/978-3-662-61313-9

[8] The power of APT attribution. [Internet]. 2016. Available from : https://media.kaspersky.com/en/business-security/enterprise/threat-attribution-engine-whitepaper.pdf. [Accessed: 25 May 2021].

[9] Rosenberg, Ishai, Guillaume Sicard, and Eli Omid David." DeepAPT: nation-state APT attribution using end-to-end deep neural networks." International Conference on Artificial Neural Networks. Springer, Cham, 2017.DOI: https://doi.org/10.1007/978-3-319-68612-7_11

[10] Perry, Lior, Bracha Shapira, and Rami Puzis." NO-DOUBT: Attack attribution based on threat intelligence reports." 2019 IEEE International Conference on Intelligence and Security Informatics (ISI). IEEE, 2019. DOI: 10.1109/ISI.2019.8823152 [Accessed: 25 May 2021]

[11] Noor, Umara, et al." A machine learning-based FinTech cyber threat attribution framework using high-level indicators of compromise." Future Generation Computer Systems 96 (2019): 227-242. DOI : https://doi.org/10.1016/j.future.2019.02.01 [Accessed: 25 May 2021]

[12] Han, Weijie, et al." APTMalInsight: Identify and cognize APT malware based on system call information and ontology knowledge framework." Information Sciences 546 (2021): 633-664. DOI: https://doi.org/10.1016/j.ins.2020.08.095 [Accessed: 25 May 2021]

[13] Going ATOMIC: Clustering and Associating Attacker Activity at Scale. [Internet]. 2019. Available from : https://www.fireeye.com/blog/threat-research/2019/03/clustering-and-associating-attacker-activity-at-scale.html [Accessed: 25 May 2021]

[14] Bodström, Tero, and Timo Hämäläinen." A novel deep learning stack for APT detection." Applied Sciences 9.6 (2019): 1055. DOI : https://doi.org/10.3390/app9061055 [Accessed: 25 May 2021]

[15] Milajerdi, Sadegh M., et al." Holmes: real-time apt detection through correlation of suspicious information flows." 2019 IEEE Symposium on Security and Privacy (SP). IEEE, 2019. DOI: 10.1109/SP.2019.00026 [Accessed: 25 May 2021]

[16] Schindler, Timo." Anomaly detection in log data using graph databases and machine learning to defend advanced persistent threats." arXiv preprint arXiv:1802.00259 (2018). DOI: 10.18420/in2017_241 [Accessed: 25 May 2021]

[17] Kim, Yong-Ho, and Won Hyung Park." A study on cyber threat prediction based on intrusion detection event for APT attack detection." Multimedia tools and applications 71.2 (2014): 685-698. DOI : https://doi.org/10.1007/s11042-012-1275-x [Accessed: 25 May 2021]

[18] Charan, PV Sai, T. Gireesh Kumar, and P. Mohan Anand." Advance persistent threat detection using long short term memory (LSTM) neural networks." International Conference on Emerging Technologies in Computer Engineering. Springer, Singapore, 2019. DOI : https://doi.org/10.1007/978-981-13-8300-7_5 [Accessed: 25 May 2021]

[19] Ghafir, Ibrahim, et al." Detection of advanced persistent threat using machine-learning correlation analysis." Future Generation Computer Systems 89 (2018): 349-359. https://doi.org/10.1016/j.future.2018.06.055 [Accessed: 25 May 2021]

[20] Siddiqui, Sana, et al." Detecting advanced persistent threats using fractal dimension based machine learning classification." Proceedings of the 2016 ACM on international workshop on security and privacy analytics. 2016. DOI: https://doi.org/10.1145/2875475.2875484 [Accessed: 25 May 2021]

[21] Giura, Paul, and Wei Wang." A context-based detection framework for advanced persistent threats." 2012 International Conference on Cyber Security. IEEE, 2012. DOI : 10.1109/CyberSecurity.2012.16 [Accessed: 25 May 2021]

[22] Fan, Chun-I., et al." Malware detection systems based on API log data mining." 2015 IEEE 39th annual computer software and applications conference. Vol. 3. IEEE, 2015. DOI : 10.1109/COMPSAC.2015.241 [Accessed: 25 May 2021]

[23] Anand, P. Mohan, T. Gireesh Kumar, and PV Sai Charan." An Ensemble approach for algorithmically generated domain name detection using statistical and lexical analysis." Procedia Computer Science 171 (2020): 1129-1136. DOI : https://doi.org/10.1016/j.procs.2020.04.121 [Accessed: 25 May 2021]

[24] Charan, PV Sai, Sandeep K. Shukla, and P. Mohan Anand." Detecting Word Based DGA Domains Using Ensemble Models." International Conference on Cryptology and Network Security. Springer, Cham, 2020. DOI : https://doi.org/10.1007/978-3-030-65411-5_7 [Accessed: 25 May 2021]

Chapter 6

Text Classification on the Instagram Caption Using Support Vector Machine

Setiawan Hadi and Paquita Putri Ramadhani

Abstract

Instagram is one of the world's top ten most popular social networks. Instagram is the most popular social networking platform in the United States, India, and Brazil, with over 1 billion monthly active users. Each of these countries has more than 91 million Instagram users. The number of Instagram users shows the various reasons and goals for them to play this social media. Social Media Marketing does not escape being one of the purposes of using Instagram, with benefits to place a market for their products. Using text classification to categorize Instagram captions into organized groups, namely fashion, food & beverage, technology, health & beauty, lifestyle & travel, this paper is expected to help people know the current trends on Instagram. The Support Vector Machine algorithm in this research is used in 66171 post captions to classify trending on Instagram. The TF-IDF (Term Frequency times Inverse Document Frequency) method and percentage variations were used for data separation in this study. This study result indicates that the use of SVM with a percentage ratio 70% of dataset for training and 30% of dataset for testing produces a higher level of accuracy compared to the others.

Keywords: Instagram, Support Vector Machine, Text Classification, TFIDF, Social Media

1. Introduction

Currently, the internet and humans cannot be separated because of the large amount of information and knowledge available on the internet with its ability to facilitate access to various things. In addition to information disclosure, the internet is also used as a place to share experiences and hobbies through social media [1].

Obtaining an overview of social media, according to Wikipedia, social media is an online platform that allows individuals to easily join, share, social networks, wikis, forums, and create blogs. Blogs, social networks, and wikis are the most common social media used by people worldwide. As of August 2017, Instagram is the sixth most popular social media platform with 700 million members.

This social media platform, commonly called IG or Insta, is an image and video sharing application that facilitates users to upload photos and videos, apply digital filters to photos and videos, and also share them on other social media [2]. Moreover, Instagram also has several other functions, namely:

1. Interact with fellow Instagram users

2. Share recommendations

3. Online marketing

4. Share hobbies or other interests

At first, social media was just a way for people to communicate with one another. As technology advances, social media allows people to express themselves as creators and thinkers, rather than just as observers. Which activities can be facilely done using Instagram. Due to the increasingly massive use of social media, marketing through social media appears to be the best option in developing their business [3].

The caption in every Instagram post is one way to attract the audience's interest to buy the goods or services being traded [4]. Audiences can interact with or respond to the post. Observations show that a post gets significantly different interactions, depending on the content of the image and the caption. When an image is uploaded with a specific caption, especially using a hashtag, the post can become a trend. The profile of a person who is a potential target market, or demographic segmentation, behavioral segmentation, and lifestyle segmentation, is related to interests. These things allow marketers to know who is paying attention and interest in the trend. According to Shopify.co.id, there are several trending Instagram categories in 2020, namely Fashion, Food & Beverage, Technology, Health & Beauty, and Lifestyle & Travel.

We can conclude that the classification of Instagram captions plays a significant role in mapping the development of trends on the platform. By knowing the latest people's favorite trends, new business people have the convenience of promoting their brand. The Instagram posts trend can be known through the text classification method. Is the trend towards Fashion, Food & Beverage, Technology, Health & Beauty, or Lifestyle & Travel? We can find out by using the Support Vector Machine algorithm.

In **Figure 1** below, the methodology used in this study is presented.

1.1 Data collections

There are two different types of datasets: data training (CSV files) and data crawling JSON as the data testing. 66.171 data are found in the data training, which contains username, caption, and labeling. There are 1.894 Instagram captions

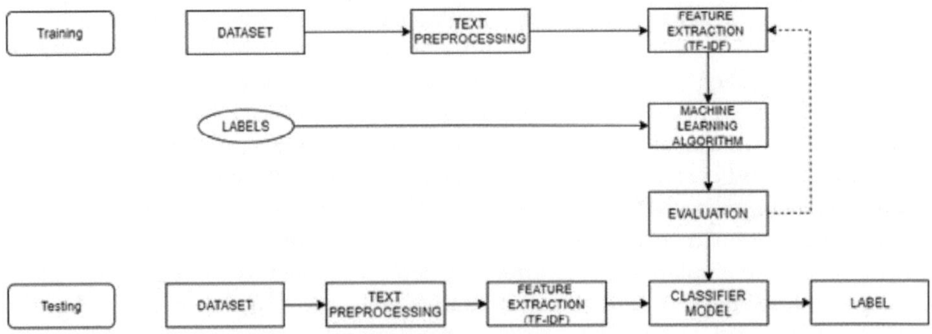

Figure 1.
Text classsification.

obtained for data testing. The caption data retrieval will be processed to produce a certain weight, which will be used later during the Instagram caption data classification process.

The data in **Table 1** is then divided into five categories; Fashion, Food & Beverage, Technology, Health & Beauty, and Lifestyle & Travel. The **Table 2** shows the proportion of the amount of data in each category:

We can see in the **Table 2** that shows an imbalanced dataset, where a disproportionate ratio is found in each class. This disproportionate ratio can be spotted in the Health & Beauty and Technology category data, which has a significant

Username	Caption	Labeling
rajvegad055	#viral #top #instatop #public #photography #editz #pose #models #look #attitude #style #bollywood #actorslife #hairstyle	1
Flexkulture	#fashion #fashionista #streetstyle #streetwearfashion #streetwear #hype #hypebeast #highfashion #offwhite #supreme #bape #balenciaga #louisvuitton #gucci #yeezy #lit #fire #drip #trending #trend #trendy #trendsetter #fashionblogger #streetstylefashion #culture #style #sneakers	1
Ishovonn	#insta #instagram #travelgram #travel #instahub #instacool #instagramhub #nature #landscape #landscapephotography #naturephotography #natureporn #ourplanetdaily #photography #fashion #streetphotography #natureaddict #naturelovers #earth #travelphotography #traveling #travelblogger #vsco #vscofilter #vscocam #canonphotographer #rainforest #cairns #australia #worl	1
Lytingz	#friendship #friends #instagood #foodie #foodlover #foodblogger #instagood #ootd #pandor #placetoeatjkt #tea #beverage	2
Willyamyantobong	#bar #band #restaurant #europeanfood #asianfood #culinary #hangout #dinner #night #friends #photooftheday #steik #foodblogger #fashionblogger #ootd #asian #asianboys #asianguys #asianwoman #candle #smile #happytummy #happy #livemusic #drink #saturdaynight #interior #song #travelblogger #weekend	2
feedmelicious	#feedmelicious #sushi #fish #sushiroll #feedme #feedmelicious #wasab #travelfood #sandiego #yummy #food #foodporn #chopsticks #tea #soysauce #tea #eatme	2
erateknologi4.0	#technology #tech #innovation #business #iphone #engineering #programming #science #design #apple #electronics #software #computer #gadgets #instagood #coding #follow #android #love #instatech #technews #geek #developer #startup #programmer #instagram #future #gadget #smartphone #bhfyp	3
tiansetia84	#technology #tech #innovation #business #iphone #engineering #electronics #science #instagood #programming #gadgets #design #art #geek #computer #coding #software #apple #android #love #smartphone #gadget #techie #developer #samsung #instatech #engineer #music #ai #bhfyp	3
ndiie_	#work #working #job #myjob #office #company #bored #grind #mygrind #dayjob #ilovemyjob #dailygrind #photooftheday #business #biz #life #workinglate #computer #instajob #instalife #instagood #instadaily	3
ayunaza69	#black #flower #pink #beauty	4
oneanonly143	#nature #beauty #liveyourlife #love	4
nagachuba_village	#fb #instagram #beauty	4
djricky07	#djricky #lifestyle #motivation #goals #entrepreneur #inspiration #busines #livelifekingsize #nightlife #fame #instagram #instafashion #instapic #photoshoot #photooftheday	5

Username	Caption	Labeling
andreigorlov	#itunes #applemusic #music #electronic #lightstorm #usa #apple #newmusic #new #travel #ipod #beats #epic #welcome #gramtrend #andreigorlov #time #apple #news #spotify #insta #love #andreigorlovofficial #photooftheday #beauty #amazing #pluto #nasa #instagood	5
athayara_	#STEPA #STEPA #STEPA #hut #smp #smpnegeri #stepa #stepamadiun #kotamadiun #madiunkota #kotagadis #instabirthday #photooftheday #LATEPOST	5

Table 1.
Instagram caption data.

Labelling	Category/Class	Data
1	Fashion	12,638
2	Food & Beverage	8,338
3	Technology	1,385
4	Health & Beauty	22,816
5	Lifestyle & Travel	20,994

Table 2.
Proportion of the amount of data In each category/class.

difference in data. This imbalanced dataset will impact the prediction process in each class later. With the imbalanced dataset, the model will tend to predict the majority class data. Meanwhile, the minority class will be treated as noise or even ignored on some occasions. Due to that, there might be misclassification of the minority class compared to the majority class. In this research, the way to resolve the imbalanced dataset is by using the performance matrix, which is the F_1 score.

1.1.1 Text preprocessing

The text preprocessing step is the beginning part of text mining. In text mining, preprocessing is the act of transforming poorly formatted input into structured data that meets the demands of the process.

The preprocessing stage is presented in **Figure 2**. After collecting the data, the next process was text processing. It included case folding, tokenizing, and cleaning.

Case folding is the process of converting the letters contained in the text into lowercase letters. Characters other than letters in the A-Z alphabet are omitted. This process was carried out due to the inconsistent use of lowercase and uppercase letters in Instagram captions. Case Folding aims to convert all data in the form of Instagram captions to conform to the standard, which usually uses lowercase letters [5]. The other characters which are not letters or numbers, like punctuation and space, will be considered as delimiter. The other characters which are not letters or numbers, like punctuation and space, will be considered as delimiter. The illustration is displayed in **Figure 3**.

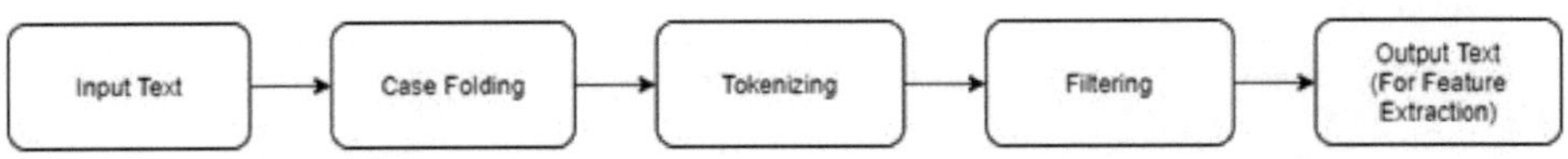

Figure 2.
Text preprocessing.

Figure 3.
Case folding process.

Figure 4.
Tokenizing process.

Figure 5.
Filtering (Stopword removal) process.

Tokenizing is a process to divides a large number of characters in a text into a single word unit by distinguishing particular characters required as a word separator [5]. Each word is identified or separated with another using space character, so this tokenizing process relies on space characters in the document to separate the words. The process is illustrated in **Figure 4**.

Filtering is a method that uses a stoplist (removing unnecessary words) or wordlist to extract certain key words from the token results (including crucial words). Some English stopword examples are "the" "from", "and", and others. The meaning behind the stopword use is to remove words with low information in a text to focus on the essential words to replace them. Filtering is done by determining what terms will be used to represent a document, where a document describes each of its contents and differs from one another. This process is illuatrated in **Figure 5**.

2. TF–IDF

The next step after text processing is TF — IDF method. At this stage, each word is assigned a weight based on how frequently it appears in the manuscript or document [6]. The computation of Term Frequency (TF) and Inverse Document Frequency (IDF) is also included in this technique (IDF). The steps are as follows:

1. Term Frequency (TF).

2. Inverse Document Frequency (IDF)

3. Term Frequency-Inverse Document Frequency (TF — IDF)

TF (Term Frequency) means the number of occurrences or the frequency of words in a document is calculated. The larger the conformity value, the more frequently a phrase appears in a text, indicating that it has a high TF [2]. Here is the formula from the TF:

$$TF = \begin{cases} 1 + \log_{10}\left(f_{t,d}\right) & , \quad f_{t,d} > 0 \\ 0 & , \quad f_{t,d} = 0 \end{cases} \tag{1}$$

A frequency term (t) in a document (d) is the value of $f_{t,d}$, d.

IDF (Inverse Document Frequency) means the distribution of a term in a collection of related texts is calculated. The relationship between the terms available in the text is also shown through this [7]. The less text a particular term contains, the larger the IDF. Here is the formula from the IDF:

$$IDF_t = \log\left({}^{N}\!/_{df_t}\right) \tag{2}$$

N: The number of text documents df_t: The total number of documents that containing the phrase "t-word" (according to the referred term).

TF — IDF is the multiplication of the results of the weighting of the frequency of a term and the frequency of the document inversely related to that term [7]. Here is the formula from the TF — IDF:

$$w_{ij} = TF \times IDF_t \tag{3}$$

TF: Term Frequency IDF_t: Inverse Document Frequency.

3. Support vector machine (SVM) model

A classification model named Support Vector Machine (SVM) is used in this method. Support Vector Machine is a supervised learning model. In its application, several linear functions of high dimensional space (feature space) are utilized. This linear function aims to find the best hyperplane in maximizing each class gap [8].

In short, support vector machine is a linear classifier. However, in some nonlinear problems, this model can also be used with some improvements [9], which are needed because not all data is linearly divided. This results in non-optimal results if linear SVM is still applied.

The radial basis function (RBF) kernel was used to change the SVM modeling process from linear to non-linear [10]. Generally, the RBF kernel is used for all types of data as a linear data separator. The RBF kernel has two parameters, namely Gamma and Cost.

The Cost parameter is used for SVM optimization so that misclassification in the training dataset sample nghwaes not occur. Meanwhile, to measure the influence given by each training dataset sample, the Gamma parameter is used [11]. A low or high value is indicated by the use of this parameter. Low or high values are described as "far" and "near". The formula below is the RBF Kernel equation:

$$K_{(x,z)} = \exp\left[-\gamma\|x - z\|^2\right] \tag{4}$$

4. F_1 score

The value of accuracy in testing the data is known by the F_1 score, which is the average of Precision and Recall, where both metrics are calculated simultaneously [10]. Precision describes the degree of precision between the required data and the model's predicted outputs [10]. The percentage of success of a model in recovering information is represented through Recall. The formula for the F_1 score is as follows:

$$F_1Score = 2 * \frac{precision * recall}{precision + recall} \tag{5}$$

The F_1 score calculation can be used as an evaluation standard from the predictive classification result if there is a class imbalance in the data.

The following are the steps taken to conduct this research:

1. The data from this study are classified into several types. Each type is labeled as follows; Fashion, Food & Beverage, Technology, Health & Beauty, and Lifestyle & Travel.

2. Case-folding and tokenizing are carried out in the data processing process by applying them to any text used in data training or data testing. After that, a new document is obtained in order to proceed to the following step.

3. At the feature extraction stage, TF-IDF is implemented in data training or data testing.

4. In the data split stage, ratios of 70:30, 60:40, 50:50, and 40:60 were used for data training.

5. This series ends with an evaluation stage in which the F1 score is used to determine the prediction results on the training data.

4.1 Result and discussion

This research begins by analyzing the dataset that has been prepared to determine whether the data has missing values, data imbalances, and other problems in the data. Proceed to the preprocessing stage to remove symbols, emoji, number punctuations, and white/multiple spaces in the text. Filtering is also done to remove words that are stop words. Then, To identify the frequency of occurrence of a word in the document, the data is transformed into vector form, and the value of Term Frequency (TF) and Inverse Document Frequency (IDF) for each token (word) is calculated. The clean and weighted data is then divided into train and testing groups with varying ratios. The support vector machine method and the radial basis function (RBF) kernel are used to classify Instagram caption data. The whole process ends with evaluating the algorithm performance using the F_1 score to overcome the imbalanced data. The difference in the distribution of train data and testing data proportion aims to see whether there is an effect of the training data and testing data proportion on the results of the F_1 score.

The outcome of the analysis is as follows:

From **Table 3**, we can see the F_1 score is obtained from the experiment. The F_1 score is generated using a distinct proportion of training and testing data and the results of Recall value and Precision. The results show that a bigger proportion of

Proportion	Precision	Recall	F_1Score
70:30	0.90	0.87	0.8895
60:40	0.90	0.86	0.8876
50:50	0.89	0.85	0.8865
40:60	0.89	0.84	0.8832

Table 3.
Comparison of precision, recall, and F_1 score for each training and testing proportion.

data training, compared to the data testing, will produce a more significant F_1 score compared to the other proportions.

Tables 4–7 show particular findings for Precision value, Recall, and F_1 Score in each category with a varied proportion of data training and testing (70:30, 60:40, 50:50, and 40:60, respectively).The following is the result of the calculation for Precision value, Recall, and F_1 Score in each data proportion:

In **Table 3** the classification results are presented using the Support Vector Machine algorithm. The average F_1 score is above 88% and the largest F_1 score is the proportion of training and testing data with a proportion of 70:30. These results are obtained through the Kernel Radial Basis Function (RBF). This proves that a larger amount of training data in a model can produce better results. The F_1 scores from each category with different training data share and testing data proportions are shown in **Tables 4–7**. The proportion of data share from training data and testing data generated is 70:30. These results are better, especially in the Technology category.

It might be interesting to split training data set and testing data set with the ratio of 80 per cent training set and 20 per cent test set and perform another experiment using that ratio. The result could give higher or lower accuracy compared with

	Precision	Recall	F_1 score
Fashion	0.83	0.82	0.83
Food & Beverage	0.92	0.87	0.90
Health & Beauty	0.89	0.95	0.92
Lifestyle & Travel	0.90	0.87	0.89
Technology	0.92	0.81	0.86

Table 4.
Proportion of data 70:30 for comparison of precision, recall, F_1 score.

Category	Precision	Recall	F_1 score
Fashion	0.83	0.82	0.83
Food & Beverage	0.92	0.86	0.89
Health & Beauty	0.90	0.95	0.92
Lifestyle & Travel	0.89	0.88	0.89
Technology	0.93	0.79	0.85

Table 5.
Proportion of 60:40 for comparison of precision, recall, F_1 score.

Category	Precision	Recall	F_1 score
Fashion	0.83	0.82	0.83
Food & Beverage	0.92	0.86	0.89
Health & Beauty	0.90	0.95	0.92
Lifestyle & Travel	0.89	0.88	0.89
Technology	0.93	0.74	0.83

Table 6.
Proportion of 50:50 for comparison of precision, recall, F_1 score.

Category	Precision	Recall	F_1 score
Fashion	0.83	0.82	0.82
Food & Beverage	0.92	0.85	0.88
Health & Beauty	0.90	0.94	0.92
Lifestyle & Travel	0.89	0.88	0.88
Technology	0.92	0.71	0.80

Table 7.
Proportion of 40:60 for comparison of precision, recall, F_1 score.

previous experiment. However, based on the references, it will be depends on the method and algorithm used.

5. Conclusion

The conclusions that can be drawn from this research are as follows:

1. In this study, a very good F_1 score, above 88%, was obtained using the Support Vector Machine (SVM) with Kernel Radial Basis Function (RBF).

2. The performance of the SVM algorithm has increased with the use of TF-IDF as a feature extraction method. The possibility of a different reaction from the algorithm, namely by not getting the expected result, can occur if there is untrained data in the data set. Data that has not been validated by experts is untrained data. Sometimes, inaccuracies can result from improper labeling of the source.

3. Model performance may be improved by dividing the data into several proportions. The use of more training also makes it possible to get better model results. This affects the researchers' use of as much data as possible to train the model.

Author details

Setiawan Hadi*† and Paquita Putri Ramadhani†
Universitas Padjadjaran, Jatinangor, Indonesia

*Address all correspondence to: setiawanhadi@unpad.ac.id

† These authors contributed equally.

References

[1] Indika D R, Jovita C: Instagram social media as a promotional tool for improving consumer buying interest. Journal of Applied Business, Polytechnic Ubaya. 2017; 1: 25–32.

[2] Chen H: College-Aged Young Consumers' Perceptions of Social Media Marketing: The Story of Instagram. Journal of Current Issues & Research in Advertising. 2017; 39: 1–15.

[3] Ting H, Ming W W P, Run E C D, Choo S L Y: Beliefs about the Use of Instagram: An Exploratory Study. International Journal of Business and Innovation. 2015: 2:15–31

[4] Adegbola O, Gearhart S., Skarda-Mitchell J.: Using Instagram to Engage with (Potential) Consumers: A Study of Forbes Most Valuable Brands' Use of Instagram The Journal of Social Media in Society. 2018: 7(2):232–251

[5] Sebastiani F: Machine Learning in Automated Text Categorization. ACM Computing Surveys. 2001: 34:1-47

[6] Kulkarni A, Shivananda A : Natural Language Processing Recipes: Unlocking Text Data with Machine Learning and Deep Learning using Python 1st ed. Edition. Apress. 2019

[7] Kedia A, Rasu M: Hands-On Python Natural Language Processing. Packt Publishing. 2020

[8] Géron A: Hands-On Machine Learning with Scikit-Learn, Keras, and TensorFlow. O'Reilly Media. 2019

[9] Steinwart I, Christmann A. Support Vector Machines. Information Science and Statistics Springer-Verlag; 2008

[10] Sokolova M, Japkowicz N, Szpakowicz S: Beyond Accuracy, F-Score and ROC: A Family of Discriminant Measures for Performance Evaluation AI. Advances in Artificial Intelligence Lecture Notes in Computer Science 4304. 2006: 1015-102

[11] Schlkopf B, Smola A. J.: Learning with Kernels: Support Vector Machines, Regularization, Optimization, and Beyond. The MIT Press. 2001

Chapter 7

Computing on Vertices in Data Mining

Leon Bobrowski

Abstract

The main challenges in data mining are related to large, multi-dimensional data sets. There is a need to develop algorithms that are precise and efficient enough to deal with big data problems. The Simplex algorithm from linear programming can be seen as an example of a successful big data problem solving tool. According to the fundamental theorem of linear programming the solution of the optimization problem can found in one of the vertices in the parameter space. The basis exchange algorithms also search for the optimal solution among finite number of the vertices in the parameter space. Basis exchange algorithms enable the design of complex layers of classifiers or predictive models based on a small number of multivariate data vectors.

Keywords: data mining, basis exchange algorithms, small samples of multivariate vectors, gene clustering, prognostic models

1. Introduction

Various data mining tools are proposed to extract patterns from data sets [1]. Large, multidimensional data sets impose high requirements as to the precision and efficiency of calculations used to extract patterns (regularities) useful in practice [2]. In this context, there is still a need to develop new algorithms of data mining [3]. New types of patterns are also obtained in result of combining different types of classification or prognosis models [4].

The Simplex algorithm from linear programming is used as an effective big data mining tool [5]. According to the basic theorem of linear programming, the solution to the linear optimization problem with linear constraints can be found at one of the vertices in the parameter space. Narrowing the search area to a finite number of vertices is a source of the efficiency of the Simplex algorithm.

Basis exchange algorithms also look for an optimal solution among a finite number of vertices in the parameter space [6]. The basis exchange algorithms are based on the Gauss - Jordan transformation and, for this reason, are similar to the Simplex algorithm. Controlling the basis exchange algorithm is related to the minimization of convex and piecewise linear (CPL) criterion functions [7].

The perceptron and collinearity criterion functions belong to the family of CPL functions The minimization of the perceptron criterion function allows to check the linear separability of data sets and to design piecewise linear classifiers [8]. Minimizing the collinearity criterion function makes it possible to detect collinear (flat) patterns in data sets and to design multiple interaction models [9].

Data sets consisting of a small number of multivariate feature vectors generate specific problems in data mining [10]. This type of data includes genetic data sets. Minimizing the perceptron criterion function or the collinearity function enables solving problems related to discrimination or regression also in the case of a small set of multidimensional feature vectors by using complex layers of low dimensional linear classifiers or prognostic models [11].

2. Linear separability vs. linear dependence

Let us assume that each of m objects O_j from a given database were represented by the n-dimensional feature vector $\mathbf{x}_j = [x_{j,1},\ldots,x_{j,n}]^T$ belonging to the feature space $F[n]$ ($\mathbf{x}_j \in F[n]$). The data set C consists of m such feature vectors $\mathbf{x}_j$:

$$C = \{\mathbf{x}_j\}, \textit{where } j = 1, \ldots, m \tag{1}$$

The components $x_{j,i}$ of the feature vector $\mathbf{x}_j$ are numerical values ($x_{j,i} \in R$ or $x_{j,I} \in \{0, 1\}$) of the individual features X_i of the j-th object O_j. In this context, each feature vector $\mathbf{x}_j$ ($\mathbf{x}_j \in F[n]$) represents n features X_i belonging to the feature set $F(n) = \{X_1, \ldots, X_n\}$.

The pairs $\{G_k^+, G_k^=\}$ ($k = 1, \ldots, K$) of the learning sets $G_k^+$and $G_k^=$ ($G_k^+ \cap G_k^- = \emptyset$) are formed from some feature vectors $\mathbf{x}_j$ selected from the data set C (1):

$$G_k^+ = \{\mathbf{x}_j : j \in J_k^+\}, \textit{and } G_k^- = \{\mathbf{x}_j : j \in J_k^-\} \tag{2}$$

where J_k^+ and J_k^- are non-empty sets of indices j of vectors $\mathbf{x}_j$ ($J_k^+ \cap J_k^- = \emptyset$).

The *positive* learning set G_k^+ is composed of m_k^+ feature vectors $\mathbf{x}_j$ ($j \in J_k^+$). Similarly, the *negative* learning set G_k^- is composed of m_k^- feature vectors $\mathbf{x}_j$ ($j \in J_k^-$), where $m_k^+ + m_k^- \le m$.

Possibility of the learning sets G_k^+ and G_k^- (2) separation using a hyperplane $H(\mathbf{w}_k, \theta_k)$ in the feature space $F[n]$ is investigated in pattern recognition [1]:

$$H(\mathbf{w}_k, \theta_k) = \{\mathbf{x} : \mathbf{w}_k^T\mathbf{x} = \theta_k\} \tag{3}$$

where $\mathbf{w}_k = [w_{k,1},\ldots, w_{k,n}]^T \in R^n$ is the weight vector, $\theta_k \in R^1$ is the threshold, and $\mathbf{w}_k^T\mathbf{x} = \Sigma_i\, w_{k,i}\, x_i$ is the scalar product.

Definition 1: The learning sets G_k^+ and G_k^- (1) are *linearly separable* in the feature space $F[n]$, if and only if there exists a weight vector $\mathbf{w}_k$ ($\mathbf{w}_k \in R^n$), and a threshold θ_k ($\theta_k \in R^1$) that the hyperplane $H(\mathbf{w}_k, \theta_k)$ (3) separates these sets [7]:

$$(\exists \mathbf{w}_k, \theta_k)\ (\forall \mathbf{x}_j \in G_k^+)\ \mathbf{w}_k^T\mathbf{x}_j \ge \theta_k + 1\ \textit{and} \tag{4}$$

$$(\forall \mathbf{x}_j \in G_k^-)\ \mathbf{w}_k^T\mathbf{x}_j \le \theta_k - 1$$

According to the above inequalities, all vectors $\mathbf{x}_j$ from the learning set G_k^+ (2) are located on the positive side of the hyperplane $H(\mathbf{w}_k, \theta_k)$ (3), and all vectors $\mathbf{x}_j$ from the set G_k^- lie on the negative side of this hyperplane.

The hyperplane $H(\mathbf{w}_k, \theta_k)$ (3) separates (4) the sets G_k^+ and G_k^- (1) with the following margin $\delta_{L2}(\mathbf{w}_k)$ based on the Euclidean (L_2) norm which is used in the Support Vector Machines (SVM) method [12]:

$$\delta_{L2}(\mathbf{w}_k) = 2/\|\mathbf{w}_k\|_{L2} = 2/(\mathbf{w}_k^T\mathbf{w}_k)^{1/2} \tag{5}$$

where $|| \mathbf{w}_k ||_{L2} = (\mathbf{w}_k^T\mathbf{w}_k)^{1/2}$ is the Euclidean length of the weight vector $\mathbf{w}_k$.

The margin $\delta_{L1}(\mathbf{w}_k)$ with the L_1 norm related to the hyperplane $H(\mathbf{w}_k, \theta_k)$ (2), which separates (10) the learning sets G_k^+ and G_k^- (2) was determined by analogy to (5) as [11]:

$$\delta_{L1}(\mathbf{w}_k) = 2/|| \mathbf{w}_k ||_{L1} = 2/(|w_{k,1}| + \ldots + | w_{k,n}|) \tag{6}$$

where $|| \mathbf{w}_k ||_{L1} = |w_{k,1}| + \ldots +| w_{k,n}|$ is the L_1 length of the weight vector $\mathbf{w}_k$.

The margins $\delta_{L2}(\mathbf{w}_k)$ (5) or $\delta_{L1}(\mathbf{w}_k)$ (6) are maximized to improve the generalization properties of linear classifiers designed from the learning sets G_k^+ and G_k^- (2) [7].

The following set of $m_k' = m_k^+ + m_k^-$ linear equations can be formulated on the basis of the linear separability inequalities (4):

$$(\forall j \in J_k^+)\ \mathbf{x}_j^T\mathbf{w}_k = \theta_k + 1\ and \tag{7}$$

$$(\forall j \in J_k^-)\ \mathbf{x}_j^T\mathbf{w}_k = \theta_k - 1$$

If we assume that the threshold θ_k can be determined latter, then we have n unknown weights $w_{k,i}$ ($\mathbf{w}_k = [w_{k,1},\ldots, w_{k,n}]^T$) in an underdetermined system of $m_k' = m_k^+ + m_k^-$ ($m_k' \le m_k < n$) linear Eqs. (7). In order to obtain a system of n linear Eqs. (7) with n unknown weights $w_{k,i}$, additional linear equations based on selected $n - m_k'$ unit vectors $\mathbf{e}_i$ ($i \in I_k$) were taken into account [6]:

$$(\forall i \in I_k)\ \mathbf{e}_i^T\mathbf{w}_k = 0 \tag{8}$$

The parameter vertex $\mathbf{w}_k = [w_{k,1},\ldots, w_{k,n}]^T$ can be determined by the linear Eqs. (7) and (8) if the feature vectors $\mathbf{x}_j$ forming the learning sets G_k^+ and G_k^- (2) are linearly independent [7].

The feature vector $\mathbf{x}_{j'}$ ($\mathbf{x}_{j'} \in G_k^+ \cup G_k^-$ (2)) is a linear combination of some other vectors $\mathbf{x}_{j(i)}$ ($j(i) \neq j'$) from the learning sets (2), if there are such parameters $\alpha_{j',i}$ ($\alpha_{j',i} \neq 0$) that the following relation holds:

$$\mathbf{x}_{j'} = \alpha_{j',1}\, \mathbf{x}_{j(1)} + \ldots + \alpha_{j',l}\, \mathbf{x}_{j(l)} \tag{9}$$

Definition 2: Feature vectors $\mathbf{x}_j$ making up the learning sets G_k^+ and G_k^- (2) are linearly independent if neither of these vectors $\mathbf{x}_{j'}$ ($\mathbf{x}_{j'} \in G_k^+ \cup G_k^-$) can be expressed as a linear combination (9) of l ($l \in \{1, \ldots, m - 1\}$) other vectors $\mathbf{x}_{j(l)}$ from the learning sets.

If the number $m_k' = m_k^+ + m_k^-$ of elements $\mathbf{x}_j$ of the learning sets G_k^+ and G_k^- (2) is smaller than the dimension n of the feature space $F[n]$ ($m_k^+ + m_k^- \le n$), then the parameter vertex $\mathbf{w}_k(\theta_k)$ can be defined by the linear equations in the following matrix form [13]:

$$B_k\, \mathbf{w}_k(\theta_k) = \mathbf{1}_k(\theta_k) \tag{10}$$

where

$$\mathbf{1}_k(\theta_k) = [\theta_k + 1, \ldots, \theta_k + 1, \theta_k - 1, \ldots, \theta_k - 1, 0, \ldots, 0]^T \tag{11}$$

and

$$B_k = \left[\mathbf{x}_1, \ldots, \mathbf{x}_{mk'}, \mathbf{e}_{i(mk'+1)}, \ldots, \mathbf{e}_{i(n)}\right]^T \tag{12}$$

The first m_k^+ components of the vector $\mathbf{1}_k(\theta_k)$ are equal to $\theta_k + 1$, the next m_k^- components equal to $\theta_k - 1$, and the last $n - m_k^+ - m_k^-$ components are equal to 0. The first m_k^+ rows of the square matrix $\boldsymbol{B}_k$ (12) are formed by the feature vectors $\mathbf{x}_j$ ($j \in J_k^+$) from the set $\boldsymbol{G}_k^+$ (2), the next m_k^- rows are formed by vectors $\mathbf{x}_j$ ($j \in J_k^-$) from the set $\boldsymbol{G}_k^-$ (2), and the last $n - m_k^+ - m_k^-$ rows are made up of unit vectors $\mathbf{e}_j$ ($i \in I_k$):

If the matrix $\boldsymbol{B}_k$ (12) is non-singular, then there exists the inverse matrix $\boldsymbol{B}_k^{-1}$:

$$\boldsymbol{B}_k^{-1} = \left[\mathbf{r}_1, \dots, \mathbf{r}_{mk'}, \mathbf{r}_{i(mk'+1)}, \dots, \mathbf{r}_{i(n)}\right] \tag{13}$$

In this case, the parameter vertex $\mathbf{w}_k(\theta_k)$ (10) can be defined by the following equation:

$$\begin{aligned}\mathbf{w}_k(\theta_k) &= \boldsymbol{B}_k^{-1}\mathbf{1}_k(\theta_k) = (\theta_k + 1)\,\mathbf{r}_k^+ + (\theta_k - 1)\,\mathbf{r}_k^- = \\ &= \theta_k\,(\mathbf{r}_k^+ + \mathbf{r}_k^-) + (\mathbf{r}_k^+ - \mathbf{r}_k^-)\end{aligned} \tag{14}$$

where the vector $\mathbf{r}_k^+$ is the sum of the first m_k^+ columns $\mathbf{r}_i$ of the inverse matrix $\boldsymbol{B}_k^{-1}$ (13), and the vector $\mathbf{r}_k^-$ is the sum of the successive m_k^- columns $\mathbf{r}_i$ of this matrix.

The last $n - (m_k^+ + m_k^-)$ components $w_{k.i}(\theta_k)$ of the vector $\mathbf{w}_k(\theta_k) = [w_{k,1}(\theta_k), \dots, w_{k,n}(\theta_k)]^T$ (14) linked to the zero components of the vector $\mathbf{1}_k(\theta_k)$ (11) are equal to zero:

$$(\forall i \in \{m_k^+ + m_k^- + 1, \dots, n\})\ w_{k.i}(\theta_k) = 0 \tag{15}$$

The conditions $w_{k.i}(\theta_k) = 0$ (15) result from the equations $\mathbf{e}_i^T\mathbf{w}_k(\theta_k) = 0$ (8) at the vertex $\mathbf{w}_k(\theta_k)$ (14).

Length $\|\mathbf{w}_k(\theta_k)\|_{L1}$ of the weight vector $\mathbf{w}_k(\theta_k)$ (14) in the L_1 norm is the sum of $m_k' = m_k^+ + m_k^-$ components $|w_{k,i}(\theta_k)|$:

$$\|\mathbf{w}_k(\theta_k)\|_{L1} = |w_{k,1}(\theta_k)| + \dots + |w_{k,mk'}(\theta_k)| \tag{16}$$

In accordance with the Eq. (14), components $|w_{k,i}(\theta_k)|$ can be determined as follows:

$$(\forall i \in \{1, \dots, m_k^+ + m_k^-\})\ |w_{k,i}(\theta_k)| = |\ \theta_k\,(r_{k,i}^+ + r_{k,i}^-) + (r_{k,i}^+ - r_{k,i}^-)\ | \tag{17}$$

The length $\|\mathbf{w}_k(\theta_k)\|_{L1}$ (16) of the vector $\mathbf{w}_k(\theta_k)$ (14) with the L_1 norm is minimized to increase the margin $\delta_{L1}(\mathbf{w}_k(\theta_k))$ (6). The length $\|\mathbf{w}_k(\theta_k)\|_{L1}$ (16) can be minimized by selecting the optimal threshold value θ_k^* on the basis of the Eq. (14).

$$(\forall \theta_k)\ \delta_{L1}(\mathbf{w}_k(\theta_k^*)) \geq \delta_{L1}(\mathbf{w}_k(\theta_k)) \tag{18}$$

where the optimal vertex $\mathbf{w}_k(\theta_k^*)$ is defined by the Eq. (14).

Theorem 1: The learning sets $\boldsymbol{G}_k^+$ and $\boldsymbol{G}_k^-$ (2) formed by m ($m \leq n$) linearly independent (9) feature vectors $\mathbf{x}_j$ are linearly separable (4) in the feature space $\boldsymbol{F}[n]$ ($\mathbf{x}_j \in \boldsymbol{F}[n]$).

Proof: If the learning sets $\boldsymbol{G}_k^+$ and $\boldsymbol{G}_k^-$ (2) are formed by m linearly independent feature.

vectors $\mathbf{x}_j$ then the non-singular matrix $\boldsymbol{B}_k = [\mathbf{x}_1, \dots, \mathbf{x}_m, \mathbf{e}_{i(m+1)}, \dots, \mathbf{e}_{i(n)}]^T$ (12) containing these m vectors $\mathbf{x}_j$ and $n - m$ unit vectors $\mathbf{e}_i$ ($i \in I_k$) can be defined [10]. In this case, the inverse matrix $\boldsymbol{B}_k^{-1}$ (13) exists and can determine the vertex $\mathbf{w}_k(\theta_k)$

(14). The vertex equation $B_k\, w_k(\theta_k) = 1_k(\theta_k)$ (10) can be reformulated for the feature vectors x_j (2) as follows:

$$(\forall x_j \in G_k^{+})\ w_k(\theta_k)^T x_j = \theta_k + 1\ and\ (\forall x_j \in G_k^{-})\ w_k(\theta_k)^T x_j = \theta_k - 1 \qquad (19)$$

The solution of the Eqs. (19) satisfies the linear separability inequalities (4).

It is possible to enlarge the learning sets G_k^+ and G_k^- (2) in such a way, which maintains their linear separability (4).

Lemma 1: Increasing the positive learning set G_k^+ (2) by such a new vector $x_{j'}$ ($x_{j'} \notin G_k^+$), which is a linear combination with the parameters $\alpha_{j',i}$ (9) of some feature vectors $x_{j(l)}$ (2) from this set ($x_{j(l)} \in G_k^+$) preserves the linear separability (4) of the learning sets if the parameters $\alpha_{j',i}$ fulfill the following condition:

$$\alpha_{j',1} + \ldots + \alpha_{j',l} \geq 1 \qquad (20)$$

If the assumptions of the lemma are met, then

$$w_k^T x_{j'} = w_k^T(\alpha_{j',1}\, x_{j(1)} + \ldots + \alpha_{j',l}\, x_{j(l)}) = \qquad (21)$$

$$= \alpha_{j',1}\, w_k^T x_{j(1)} + \ldots + \alpha_{j',l}\, w_k^T x_{j(l)} = \alpha_{j',1}\,(\theta_k + 1) + \ldots + \alpha_{j',l}\,(\theta_k + 1) \geq \theta_k + 1$$

The above inequality means that linear separability connditions (4) still apply after the increasing of the learning set G_k^+ (2).

Lemma 2: Increasing the negative learning set G_k^- (2) by such a new vector $x_{j'}$ ($x_{j'} \notin G_k^-$), which is a linear combination with the parameters $\alpha_{j',i}$ (9) of some feature vectors $x_{j(l)}$ (2) from this set ($x_{j(l)} \in G_k^-$) preserves the linear separability (4) of the learning sets if the parameters $\alpha_{j',\,i}$ fulfill the following condition:

$$\alpha_{j',1} + \ldots + \alpha_{j',l} \leq -1 \qquad (22)$$

Justification Lemma 2 may be based on inequality similar to (21).

3. Perceptron criterion function

The minimization the perceptron criterion function allows to assess the degree of linear separabilty (4) of the learning sets G_k^+ and G_k^- (2) in different feature subspaces $F[n']$ ($F[n'] \subset F[n + 1]$) [6]. When defining the perceptron criterion function, it is convenient to use the following augmented feature vectors y_j ($y_j \in F[n + 1]$) and augmented weight vectors v_k ($v_k \in R^{n+1}$) [1]:

$$(\forall j \in J_k^+\ (2))\ y_j = [x_j^T, 1]^T, \qquad (23)$$

$$(\forall j \in J_k^-(2))\ y_j = -[x_j^T, 1]^T$$

and

$$v_k = [w_k^T, -\theta_k]^T = [w_{k,1}, \ldots, w_{k,n}, -\theta_k]^T \qquad (24)$$

The augmented vectors y_j are constructed (23) on the basis of the learning sets G_k^+ and G_k^- (2). These learning sets are extracted from the data set C (1) according to some additional knowledge. The linear separability (4) of the learning sets G_k^+

and G_k^- (2) can be reformulated using the following set of m inequalities with the augmented vectors $\mathbf{y}_j$ (23) [7]:

$$(\exists \mathbf{v}_k)\ (\forall j \in J_k^+ \cup J_k^-\ (2))\ \mathbf{v}_k^T \mathbf{y}_j \geq 1 \tag{25}$$

The dual hyperplanes h_j^P in the parameter space R^{n+1} ($\mathbf{v} \in R^{n+1}$) are defined on the basis of the augmented vectors $\mathbf{y}_j$ [6]:

$$(\forall j \in J_k^+ \cup J_k^-\ (2))\ h_j^P = \{\mathbf{v} : \mathbf{y}_j^T \mathbf{v} = 1\} \tag{26}$$

Dual hyperplanes h_j^P (26) divide the parameter space R^{n+1} ($\mathbf{v} \in R^{n+1}$) into a finite number L of disconnected regions (*convex polyhedra*) D_l^P ($l = 1, \dots, L$) [7]:

$$D_l^P = \{\mathbf{v} : (\forall j \in J_l^+)\ \mathbf{y}_j^T \mathbf{v} \geq 1\ \mathit{and}\ (\forall j \in J_l^-)\ \mathbf{y}_j^T \mathbf{v} < 1\} \tag{27}$$

where J_l^+ and J_l^- are disjointed subsets ($J_l^+ \cap J_l^+ = \varnothing$) of indices j of feature vectors $\mathbf{x}_j$ making up the learning sets G_k^+ and G_k^- (2).

The perceptron penalty functions $\varphi_j^P(\mathbf{v})$ are defined as follows for each of augmented feature vectors $\mathbf{y}_j$ (23) [6]:

$$(\forall j \in J_k)\quad \varphi_j^P(\mathbf{v}) = \begin{cases} 1 - \mathbf{y}_j^T \mathbf{v} & \mathit{if}\ \ \mathbf{y}_j^T \mathbf{v} < 1 \\ 0 & \mathit{if}\ \ \mathbf{y}_j^T \mathbf{v} \geq 1 \end{cases} \tag{28}$$

The j - th penalty function $\varphi_j^P(\mathbf{v})$ (28) is greater than zero if and only if the weight vector $\mathbf{v}$ is located on the wrong side ($\mathbf{y}_j^T \mathbf{v} < 1$) of the j-th dual hyperplane h_j^P (26). The function $\varphi_j^P(\mathbf{v})$ (28) is linear and greater than zero as long as the parameter vector $\mathbf{v} = [v_{k,1}, \dots, v_{k,n+1}]^T$ remains on the wrong side of the hyperplane h_j^P (26). Convex and piecewise-linear (*CPL*) penalty functions $\varphi_j^P(\mathbf{v})$ (28) are used to enforce the linear separation (8) of the learning sets G_k^+ and G_k^- (2).

The perceptron criterion function $\Phi_k^P(\mathbf{v})$ is defined as the weighted sum of the penalty functions $\varphi_j^P(\mathbf{v})$ (28) [6]:

$$\Phi_k^P(\mathbf{v}) = \Sigma_j\ \alpha_j\ \varphi_j^P(\mathbf{v}) \tag{29}$$

Positive parameters α_j ($\alpha_j > 0$) can be treated as prices of individual feature vectors $\mathbf{x}_j$:

$$(\forall j \in J_k^+\ (2))\ \alpha_j = 1/(2\, m_k^+)\ \mathit{and}\ (\forall j \in J_k^-\ (2))\ \alpha_j = 1/(2\, m_k^-) \tag{30}$$

where m_k^+ (m_k^-) is the number of elements $\mathbf{x}_j$ in the learning set G_k^+ (G_k^-) (2).

The perceptron criterion function $\Phi_k^P(\mathbf{v})$ (29) was built on the basis of the *error correction* algorithm, the basic algorithm in the *Perceptron* model of learning processes in neural networks [14].

The criterion function $\Phi_k^P(\mathbf{v})$ (29) is convex and piecewise-linear (*CPL*) [6]. It means, among others, that the function $\Phi_k^P(\mathbf{v})$ (29) remains linear within each area D_l (27):

$$(\forall l \in \{1, \dots, L\})\quad (\forall \mathbf{v} \in D_l)\ \Phi_k^P(\mathbf{v}) = \left(\Sigma_j\ \alpha_j\ \mathbf{y}_j\right)^T \tag{31}$$

where the summation is performed on all vectors $\mathbf{y}_j$ (23) fulfilling the condition $\mathbf{y}_j^T\mathbf{v} < 1$.

The optimal vector $\mathbf{v}_k^*$ determines the minimum value $\Phi_k^P(\mathbf{v}_k^*)$ of the criterion function $\Phi_k^P(\mathbf{v})$ (29):

$$(\exists \mathbf{v}_k^*)\ (\forall \mathbf{v} \in \mathbf{R}^{n+1})\ \Phi_k^P(\mathbf{v}) \geq \Phi_k^P(\mathbf{v}_k^*) \geq 0 \tag{32}$$

Since the criterion function $\Phi_k^P(\mathbf{v})$ (29) is linear in each convex polyhedron $\mathbf{D}_l$ (27), the optimal point $\mathbf{v}_k^*$ representing the minimum $\Phi_k^P(\mathbf{v}_k^*)$ (32) can be located in selected vertex of some polyhedron $\mathbf{D}_{l'}^P$ (27). This property of the optimal vector $\mathbf{v}_k^*$ (32) follows from the *fundamental theorem of linear programming* [5].

It has been shown that the minimum value $\Phi_k^P(\mathbf{v}_k^*)$ (32) of the perceptron criterion function $\Phi_k^P(\mathbf{v})$ (29) with the parameters α_j (30) is normalized as follows [6]:

$$0 \leq \Phi_k^P(\mathbf{v}_k^*) \leq 1 \tag{33}$$

The below theorem has been proved [6]:

Theorem 2: The minimum value $\Phi_k^P(\mathbf{v}_k^*)$ (32) of the perceptron criterion function $\Phi_k^P(\mathbf{v})$ (29) is equal to zero ($\Phi_k^P(\mathbf{v}_k^*) = 0$) if and only if the learning sets $\mathbf{G}_k^+$ and $\mathbf{G}_k^-$ (2) are linearly separable (4).

The minimum value $\Phi_k^P(\mathbf{v}_k^*)$ (32) is near to one ($\Phi_k^P(\mathbf{v}_k^*) \approx 1$) if the sets $\mathbf{G}_k^+$ and $\mathbf{G}_k^-$ (2) cover almost completely. It can also be proved that the minimum value $\Phi_k^P(\mathbf{v}_k^*)$ (32) of the perceptron criterion function $\Phi_k^P(\mathbf{v})$ (29) does not depend on invertible linear transformations of the feature vectors $\mathbf{y}_j$ (23) [6]. The perceptron criterion function $\Phi_k(\mathbf{v})$ (29) remains linear inside of each region $\mathbf{D}_l$ (27).

The regularized criterion function $\Psi_k^P(\mathbf{v})$ is defined as the sum of the perceptron criterion function $\Phi_k^P(\mathbf{v})$ (29) and some additional penalty functions [13]. These additional *CPL* functions are equal to the costs γ_i ($\gamma_i > 0$) of individual features X_i multiply by the absolute values $|w_i|$ of weighs w_i, where $\mathbf{v} = [\mathbf{w}^T, -\theta]^T = [w_1,..., w_n, -\theta]^T \in \mathbf{R}^{n+1}$ (24):

$$\Psi_k^P(\mathbf{v}) = \Phi_k^P(\mathbf{v}) + \lambda\ \Sigma_i\ \gamma_i|w_i| \tag{34}$$

where λ ($\lambda \geq 0$) is the *cost level*. The standard values of the cost parameters γ_i are equal to one ($\forall i \in \{1, ..., n\}\ \gamma_i = 1$).

The optimal vector $\mathbf{v}_{k,\lambda}^*$ constitutes the minimum value $\Psi_k^P(\mathbf{v}_{k,\lambda}^*)$ of the *CPL* criterion function $\Psi_k^P(\mathbf{v})$ (34), which is defined on elements $\mathbf{x}_j$ of the learning sets $\mathbf{G}_k^+$ and $\mathbf{G}_k^-$ (2):

$$(\exists \mathbf{v}_{k,\lambda}^*)\ (\forall \mathbf{v} \in R^{n+1})\ \Psi_k^P(\mathbf{v}) \geq \Psi_k^P(\mathbf{v}_{k,\lambda}^*) > 0 \tag{35}$$

Similarly as in the case of the perceptron criterion function $\Phi_k^P(\mathbf{v})$ (29), the optimal vector $\mathbf{v}_{k,\lambda}^*$ (35) can be located in selected vertex of some polyhedron $\mathbf{D}_{l'}$ (27). The minimum value $\Psi_k^P(\mathbf{v}_{k,\lambda}^*)$ (35) of the criterion function $\Psi_k^P(\mathbf{v})$ (34) is used, among others, in the *relaxed linear separability* (*RLS*) method of gene subsets selection [15].

4. Collinearity criterion function

Minimizing the collinearity criterion function is used to extract collinear patterns from large, multidimensional data sets $\mathbf{C}$ (1) [7]. Linear models of multivariate interactions can be formulated on the basis of representative collinear patterns [9].

The collinearity penalty functions $\varphi_j(\mathbf{w})$ are determined by individual feature vectors $\mathbf{x}_j = [x_{j,1},\ldots,x_{j,n}]^T$ in the following manner [9]:

$$(\forall \mathbf{x}_j \in \boldsymbol{C}\ (1))$$
$$\varphi_j(\mathbf{w}) = |\,1 - \mathbf{x}_j^T\mathbf{w}\,| = \begin{cases} 1 - \mathbf{x}_j^T\mathbf{w} & \textit{if}\ \ \mathbf{x}_j^T\mathbf{w} \le 1 \\ \mathbf{x}_j^T\mathbf{w} - 1 & \textit{if}\ \ \mathbf{x}_j^T\mathbf{w} > 1 \end{cases} \quad (36)$$

The penalty functions $\varphi_j(\mathbf{w})$ (36) can be related to the following dual hyperplanes h_j^1 in the parameter (weight) space $\boldsymbol{R}^n$ ($\mathbf{w} \in \boldsymbol{R}^n$):

$$(\forall j = 1, \ldots, m)\ h_j^1 = \{\mathbf{w} : \mathbf{x}_j^T\mathbf{w} = 1\} \quad (37)$$

The *CPL* penalty $\varphi_j(\mathbf{w})$ (36) is equal to zero ($\varphi_j^c(\mathbf{w}) = 0$) in the point $\mathbf{w} = [w_1,\ldots, w_n]^T$ if and only if the point $\mathbf{w}$ is located on the dual hyperplane h_j^1 (37).

The collinearity criterion function $\Phi_k(\mathbf{w})$ is defined as the weighted sum of the penalty functions $\varphi_j(\mathbf{w})$ (36) determined by feature vectors $\mathbf{x}_j$ forming the data subset $\boldsymbol{C}_k$ ($\boldsymbol{C}_k \subset \boldsymbol{C}$ (1)):

$$\Phi_k(\mathbf{w}) = \Sigma_j\ \beta_j\ \varphi_j(\mathbf{w}) \quad (38)$$

where the sum takes into account only the indices J of the set $J_k = \{j: \mathbf{x}_j \in \boldsymbol{C}_k\}$, and the positive parameters β_j ($\beta_j > 0$) in the function $\Phi_k(\mathbf{w})$ (38) can be treated as the *prices* of particular feature vectors $\mathbf{x}_j$. The standard choice of the parameters β_j values is one ($(\forall j \in J_k)\ \beta_j = 1.0$).

The collinearity criterion function $\Phi_k(\mathbf{w})$ (38) is convex and piecewise-linear (*CPL*) as the sum of this type of penalty functions $\varphi_j(\mathbf{w})$ (36) [9]. The vector $\mathbf{w}_k^*$ determines the minimum value $\Phi_k(\mathbf{w}_k^*)$ of the criterion function $\Phi_k(\mathbf{w})$ (38):

$$(\exists \mathbf{w}_k^*)\ (\forall \mathbf{w})\ \Phi_k(\mathbf{w}) \ge \Phi_k(\mathbf{w}_k^*) \ge 0 \quad (39)$$

Definition 3: The data subset $\boldsymbol{C}_k$ ($\boldsymbol{C}_k \subset \boldsymbol{C}$ (1)) is *collinear* when all feature vectors $\mathbf{x}_j$ from this subset are located on some hyperplane $H(\mathbf{w}, \theta) = \{\mathbf{x}: \mathbf{w}^T\mathbf{x} = \theta\}$ with $\theta \neq 0$.

Theorem 3: The minimum value $\Phi_k^P(\mathbf{v}_k^*)$ (39) of the collinearity criterion function $\Phi_k(\mathbf{w})$ (38) defined on the feature vectors $\mathbf{x}_j$ constituting a data subset $\boldsymbol{C}_k$ ($\boldsymbol{C}_k \subset \boldsymbol{C}$ (1)) is equal to zero ($\Phi_k^P(\mathbf{v}_k^*) = 0$) when this subset $\boldsymbol{C}_k$ is collinear (*Def.* 3) [9].

Different collinear subsets $\boldsymbol{C}_k$ can be extracted from data set $\boldsymbol{C}$ (1) with a large number m of elements $\mathbf{x}_j$ by minimizing the collinearity criterion function $\Phi_k^P(\mathbf{w})$ (38) [9].

The minimum value $\Phi_k^P(\mathbf{v}_k^*)$ (39) of the collinearity criterion function $\Phi_k(\mathbf{w})$ (38) can be reduced to zero by omitting some feature vectors $\mathbf{x}_j$ from the data subset $\boldsymbol{C}_k$ ($\boldsymbol{C}_k \subset \boldsymbol{C}$ (1)). If the minimum value $\Phi_k(\mathbf{w}_k^*)$ (39) is greater than zero ($\Phi_k(\mathbf{w}_k^*) > 0$) then we can select feature vectors $\mathbf{x}_j$ ($j \in J_k(\mathbf{w}_k^*)$) with the penalty $\varphi_j(\mathbf{w}_k^*)$ (36) greater than zero:

$$(\forall j \in J_k(\mathbf{w}_k^*))\ \varphi_j(\mathbf{w}_k^*) = |1 - \mathbf{x}_j^T\ \mathbf{w}_k^*| > 0 \quad (40)$$

Omitting one feature vector $\mathbf{x}_{j'}$ ($j' \in J_k(\mathbf{w}_k^*)$) with the above property results in the following reduction of the minimum value $\Phi_k^P(\mathbf{v}_k^*)$ (39);

$$\Phi_{k'}(\mathbf{w}_{k'}^*) \le \Phi_k(\mathbf{w}_k^*) - \varphi_{j'}(\mathbf{w}_k^*) \quad (41)$$

where $\Phi_{k'}(\mathbf{w}_{k'}^*)$ is the minimum value (39) of the collinearity criterion function $\Phi_{k'}(\mathbf{w})$ (38) defined on feature vectors $\mathbf{x}_j$ constituting the data subset C_k reduced by the vector $\mathbf{x}_{j'}$.

The regularized criterion function $\Psi_k(\mathbf{w})$ is defined as the sum of the collinearity criterion function $\Phi_k(\mathbf{w})$ (38) and some additional *CPL* penalty functions $\varphi_j^0(\mathbf{w})$ [7]:

$$\Psi_k(\mathbf{w}) = \Phi_k(\mathbf{w}) + \lambda\, \Sigma_i\, \chi_i(\mathbf{w}) = \Sigma_j\, \beta_j\, \varphi_j(\mathbf{w}) + \lambda\, \Sigma_i\, \chi_i\, \varphi_i^0(\mathbf{w}) \qquad (42)$$

where $\lambda \geq 0$ is the *cost level*. The standard values of the cost parameters γ_i are equal to one $((\forall i \in \{1, \dots, n\})\ \gamma_i = 1)$. The additional *CPL* penalty functions $\varphi_j^0(\mathbf{w})$ are defined below [7]:

$$(\forall i = 1, \dots, n) \qquad (43)$$

$$\chi_i(\mathbf{w}) = |\, \mathbf{e}_i^T \mathbf{w}\, | = \begin{matrix} -w_j & if & w_j \leq 0 \\ w_j & if & w_j > 0 \end{matrix}$$

The functions $\varphi_j^0(\mathbf{w})$ (43) are related to the following dual hyperplanes h_j^0 in the parameter (*weight*) space R^n ($\mathbf{w} \in R^n$):

$$(\forall i = 1, \dots, n)\ h_j^0 = \{\mathbf{w} : \mathbf{e}_j^T \mathbf{w} = 0\} = \{\mathbf{w} : \mathbf{w}_j = 0\} \qquad (44)$$

The *CPL* penalty function $\varphi_j^0(\mathbf{w})$ (43) is equal to zero ($\varphi_j^0(\mathbf{w}) = 0$) in the point $\mathbf{w} = [w_1, \dots, w_n]^T$ if and only if this point is located on the dual hyperplane h_j^0 (44).

5. Parameter vertices

The perceptron criterion function $\Phi_k^P(\mathbf{v})$ (29) and the collinearity criterion function $\Phi_k(\mathbf{w})$ (38) are convex and piecewise-linear (*CPL*). The minimum values of a such *CPL* criterion functions can be located in parameter vertices of some convex polyhedra. We consider the parameter vertices $\mathbf{w}_k$ ($\mathbf{w}_k \in R^n$) related to the collinearity criterion function $\Phi_k(\mathbf{w})$ (38).

Definition 4: The *parameter vertex* $\mathbf{w}_k$ of the *rank* r_k ($r_k \leq n$) in the weight space R^n ($\mathbf{w}_k \in R^n$) is the intersection point of r_k hyperplanes h_j^1 (37) defined by linearly indepenedent feature vectors $\mathbf{x}_j$ ($j \in J_k$) from the data set C (1) and $n - r_k$ hyperplanes h_i^0 (44) defined by unit vectors $\mathbf{e}_i$ ($i \in I_k$) [7].

The j-th dual hyperplane h_j^1 (37) defined by the feature vector $\mathbf{x}_j$ (1) passes through the k-th *vertex* $\mathbf{w}_k$ if the equation $\mathbf{w}_k^T \mathbf{x}_j = 1$ holds.

Definition 5: The k-th weight vertex $\mathbf{w}_k$ of the rank r_k is *degenerate* in the parameter space R^n if the number m_k of hyperplanes h_j^1 (37) passing through this vertex ($\mathbf{w}_k^T \mathbf{x}_j = 1$) is greater than the rank r_k ($m_k > r_k$).

The vertex $\mathbf{w}_k$ can be defined by the following set of n linear equations:

$$(\forall j \in J_k(\mathbf{w}_k))\ \mathbf{w}_k^T \mathbf{x}_j = 1 \qquad (45)$$

and

$$(\forall i \in I_k(\mathbf{w}_k))\ \mathbf{w}_k^T \mathbf{e}_i = 0 \qquad (46)$$

Eqs. (45) and (46) can be represented in the below matrix form [7]:

$$\mathbf{B}_k\, \mathbf{w}_k = \mathbf{1}_k \qquad (47)$$

where $\mathbf{1}_k = [1, \ldots, 1, 0, \ldots, 0]^T$ is the vector with the first r_k components equal to one and the remaining $n - r_k$ components are equal to zero.

The square matrix B_k (47) consists of k feature vectors $\mathbf{x}_j$ ($j \in J_k$ (45)) and $n - k$ unit vectors $\mathbf{e}_i$ ($i \in I_k$ (46)) []:

$$B_k = \left[\mathbf{x}_1, \ldots, \mathbf{x}_k, \mathbf{e}_{i(k+1)}, \ldots, \mathbf{e}_{i(n)}\right]^T \tag{48}$$

where the symbol $\mathbf{e}_{i(l)}$ denotes such unit vector, which is the l-th row of the matrix B_k.

Since feature vectors $\mathbf{x}_j$ ($\forall j \in J_k(\mathbf{w}_k)$ (45)) making up r_k rows of the matrix B_k (48) are linearly independent, then the inverse matrix B_k^{-1} exists:

$$B_k^{-1} = \left[\mathbf{r}_1, \ldots, \mathbf{r}_k, \mathbf{r}_{i(k+1)}, \ldots, \mathbf{r}_{i(n)}\right] \tag{49}$$

The inverse matrix B_k^{-1} (49) can be obtained starting from the unit matrix $I = [\mathbf{e}_1, \ldots, \mathbf{e}_n]^T$ and using the basis exchange algorithm [8].

The non-singular matrix $\mathbf{B}_k$ (48) is the *basis* of the feature space $F[n]$ related to the vertex $\mathbf{w}_k = [w_{k,1}, \ldots, w_{k,n}]^T$. Since the last $n - r_k$ components of the vector $\mathbf{1}_k$ (47) are equal to zero, the following equation holds:

$$\mathbf{w}_k = B_k^{-1}\mathbf{1}_k = \mathbf{r}_1 + \ldots + \mathbf{r}_k \tag{50}$$

According to Eq. (50), the weight vertex $\mathbf{w}_k$ is the sum of the first k columns $\mathbf{r}_i$ of the inverse matrix B_k^{-1} (49).

Remark 1: The $n - k$ components $w_{k.i}$ of the vector $\mathbf{w}_k = [w_{k,1}, \ldots, w_{k,n}]^T$ (50) linked to the zero components of the vector $\mathbf{1}_k = [1, \ldots, 1, 0, \ldots, 0, 1]^T$ (7) are equal to zero:

$$(\forall i \in \{k+1, \ldots, n\})\ w_{k.i} = 0 \tag{51}$$

The conditions $w_{k.i} = 0$ (51) result from the equations $\mathbf{w}_k^T\mathbf{e}_i = 0$ (46) at the vertex $\mathbf{w}_k$.

The *fundamental theorem of linear programming* shows that the minimum $\Phi_k(\mathbf{w}_k^*)$ (39) of the *CPL* collinearity criterion function $\Phi_k(\mathbf{w})$ (38) can always be located in one of the vertices $\mathbf{w}_k$ (50) [5]. The same property has also the regularized criterion function $\Psi_k(\mathbf{w})$ (42), another function of the *CPL* type [7].

We can see that all such feature vectors $\mathbf{x}_j$ (1) which define hyperplanes h_j^1 (37) passing through the vertex $\mathbf{w}_k$ are located on the hyperplane $H(\mathbf{w}_k, 1) = \{\mathbf{x}: \mathbf{w}_k^T\mathbf{x} = 1\}$ (3) in the feature space $F[n]$. A large number m_k of feature vectors $\mathbf{x}_j$ (1) located on the hyperplane $H(\mathbf{w}_k, 1)$ (3) form the *collinear cluster* $C(\mathbf{w}_k)$ based on the vertex $\mathbf{w}_k$ [8]:

$$C(\mathbf{w}_k) = \left\{\mathbf{x}_j \in C\ (1) : \mathbf{w}_k^T\mathbf{x} = 1\right\} \tag{52}$$

If the vertex $\mathbf{w}_k$ of the rank r_k is degenerate in the parameter space R^n then the collinear cluster $C(\mathbf{w}_k)$ (52) contains more than r_k feature vectors $\mathbf{x}_j$ (1).

The k-th vertex $\mathbf{w}_k = [w_{k,1}, \ldots, w_{k,n}]^T$ in the parameter space R^n ($\mathbf{w}_k \in R^n$) is linked by the Eq. (47) to the non-singular matrix $\mathbf{B}_k$ (48). The rows of the matrix $\mathbf{B}_k$ (48) can form the *basis* of the feature space $F[n]$. The conditions $w_{k.i} = 0$ (51) result from the equations $\mathbf{w}_k^T\mathbf{e}_i = 0$ (46) at the vertex $\mathbf{w}_k$.

$$(\forall i = 1, \ldots, n)\ \textit{if}\ (\mathbf{e}_i \in \mathbf{B}_k\ (48)), \textit{then}\ w_{k.i} = 0 \tag{53}$$

Each feature vector $\mathbf{x}_j$ from the data set C (1) represents n features X_i belonging to the feature set $R(n) = \{X_1, \ldots, X_n\}$. The k-th *vertexical feature subset* $R_k(r_k)$ consists of r_k features X_i that are connected to the weights $w_{k.i}$ different from zero ($w_{k.i} \neq 0$):

$$R_k(r_k) = \{X_{i(1)}, \ldots, X_{i(rk)}\} \tag{54}$$

The k-th *vertexical subspace* $F_k[r_k]$ ($F_k[r_k] \subset F[n]$) contains the reduced vectors $\mathbf{x}_j[r_k]$ with r_k componets $x_{j,i(l)}$ ($\mathbf{x}_j[r_k] \in F_k[r_k]$) related to the weights $w_{k.i}$ different from zero:

$$(\forall j \in \{1, \ldots, m\}) \ \mathbf{x}_j[r_k] = [x_{j,i(1)}, \ldots, x_{j,i(rk)}]^T \tag{55}$$

The reduced vectors $\mathbf{x}_j[r_k]$ (55) are obtained from the feature vectors $\mathbf{x}_j = [x_{j,1},\ldots,x_{j,n}]^T$ belonging to the data set C (1) by omitting the $n - r_k$ components $x_{j,i}$ related to the weights $w_{k.i}$ equal to zero ($w_{k.i} = 0$).

We consider the optimal vertexical subspace $F_k^*[r_k]$ ($F_k^*[r_k] \subset F[n]$) related to the reduced optimal vertex $\mathbf{w}_k^*[r_k]$ which determines the minimum $\Phi_k(\mathbf{w}_k^*)$ (39) of the collinearity criterion function $\Phi_k(\mathbf{w})$ (38). The optimal collinear cluster $C(\mathbf{w}_k^*[r_k])$ (52) is based on the optimal vertex $\mathbf{w}_k^*[r_k] = [w_{k,1}^*, \ldots, w_{k,rk}^*]^T$ with r_k different from zero components $w_{k,i}^*$ ($w_{k.i}^* \neq 0$). Feature vectors $\mathbf{x}_j$ belonging to the collinear cluster $C(\mathbf{w}_k^*)$ (52) satisfy the equations $\mathbf{w}_k^*[r_k]^T\mathbf{x}_j[r_k] = 1$, hence:

$$\begin{aligned}&(\forall \mathbf{x}_j \in P(\mathbf{w}_k^*))\\ &w_{k.1}^* \ x_{j,i(1)} + \ldots + w_{k.rk}^* \ x_{j,i(rk)} = 1\end{aligned} \tag{56}$$

where $x_{j,i(l)}$ are components of the j-th feature vectors $\mathbf{x}_j$ related to the weights $w_{k.i}$ different from zero ($w_{k.i} \neq 0$).

A large number m_k of feature vectors $\mathbf{x}_j$ (1) belonging to the collinear cluster $C(\mathbf{w}_k^*[r_k])$ (52) justifies the following collinear model of interaction between selected features $X_{i(l)}$ which is based on the Eqs. (56) [9]:

$$w_{k.1}^* \ X_{i(1)} + \ldots + w_{k.rk}^* \ X_{i(rk)} = 1 \tag{57}$$

The collinear interaction model (57) allows, inter alia, to design the following prognostic models for each feature $X_{i'}$ from the subset $R_k(r_k)$ (54):

$$(\forall i' \in \{1, \ldots, r_k\}) \ X_{i'} = \alpha_{i',0} + \alpha_{i',1} X_{i(1)} + \ldots + \alpha_{i',rk} X_{i(rk)} \tag{58}$$

where $\beta_{i',0} = 1 / w_{k.i'}^*$, $\beta_{i',i'} = 0$, and ($\forall\, i(l) \neq i'$) $\beta_{i',i(l)} = w_{k.i(l)}^* / w_{k.i'}^*$.

Feature $X_{i'}$ is a dependent variable in the prognostic model (58), the remaining $m - 1$ features $X_{i(l)}$ are independent variables ($i(l) \neq i'$). The family of r_k prognostic models (58) can be designed on the basis of one collinear interaction model (57). Models (58) have a better justification for a large number m_k of feature vectors $\mathbf{x}_j$ (1) in the collinear cluster $C(\mathbf{w}_k^*[r_k])$ (52).

6. Basis exchange algorithm

The collinearity criterion function $\Phi(\mathbf{w})$ (38), like other convex and piecewise linear (*CPL*) criterion functions, can be minimized using the basis exchange algorithm [8]. The basis exchange algorithm aimed at minimization of the collinearity criterion function $\Phi(\mathbf{w})$ (38) is described below.

According to the basis exchange algorithm, the optimal vertex $\mathbf{w}_k^*$, which constitutes the minimum value $\Phi_k(\mathbf{w}_k^*)$ (39) of the collinearity function $\Phi_k(\mathbf{w})$ (38), is achieved after a finite number L of the steps l as a result of guided movement between selected vertices $\mathbf{w}_k$ (50) [8]:

$$\mathbf{w}_0 \rightarrow \mathbf{w}_1 \rightarrow \ldots . \rightarrow \mathbf{w}_L \tag{59}$$

The sequence of vertices $\mathbf{w}_k$ (59) is related by (47) to the following sequence of the inverse matrices $\boldsymbol{B}_k^{-1}$ (49):

$$\boldsymbol{B}_0^{-1} \rightarrow \boldsymbol{B}_1^{-1} \rightarrow \ldots . \rightarrow \boldsymbol{B}_L^{-1} \tag{60}$$

The sequence of vertices $\mathbf{w}_{k(l)}$ (59) typically starts at the vertex $\mathbf{w}_0 = [0,\ldots, 0]^T$ related to the identity matrix $\boldsymbol{B}_0 = \boldsymbol{I}_n = [\mathbf{e}_1,\ldots, \mathbf{e}_n]^T$ of the dimension n x n [7]. The final vertex $\mathbf{w}_L$ (59) should assure the minimum value of the collinearity criterion function $\Phi(\mathbf{w})$ (38):

$$(\forall \mathbf{w})\ \Phi(\mathbf{w}) \geq \Phi(\mathbf{w}_L) \geq 0 \tag{61}$$

If the criterion function $\Phi(\mathbf{w})$ (38) is defined on m ($m \leq n$) linearly independent vectors $\mathbf{x}_j$ ($\mathbf{x}_j \in \boldsymbol{C}$ (1)) then the value $\Phi(\mathbf{w}_L)$ of the collinearity criterion function $\Phi(\mathbf{w})$ (38) at the final vertex $\mathbf{w}_L$ (59) becomes zero ($\Phi(\mathbf{w}_L) = 0$) [8]. The rank r_L (*Def.* 4) of the final vertex $\mathbf{w}_L$ (59) can be equal to the number m of feature vectors $\mathbf{x}_j$ ($r_L = m$) or it can be less than m ($r_L < m$). The rank r_L of the final vertex $\mathbf{w}_L$ (59) is less than m ($r_L < m$) if the final vertex $\mathbf{w}_L$ is degenerate [7].

Consider the reversible matrix $\boldsymbol{B}_k = [\mathbf{x}_1,\ldots, \mathbf{x}_k, \mathbf{e}_{i(k+1)},\ldots, \mathbf{e}_{i(n)}]^T$ (48), which determines the vertex $\mathbf{w}_k$ (50) and the value $\Phi_k(\mathbf{w}_k)$ of the criterion function $\Phi_k(\mathbf{w})$ (38) in the k-th step. In the step (l + 1), one of the unit vectors $\mathbf{e}_i$ in the matrix $\mathbf{B}_k$ (48) is replaced by the feature vector $\mathbf{x}_{k+1}$ and the matrix $\mathbf{B}_{k+1} = [\mathbf{x}_1,\ldots, \mathbf{x}_k, \mathbf{x}_{k+1}, \mathbf{e}_{i(k+2)},\ldots, \mathbf{e}_{i(n)}]^T$ appears. The unit vector $\mathbf{e}_{i(k+1)}$ leaving matrix $\boldsymbol{B}_k$ (48) is indicated by an *exit criterion* based on the gradient of the collinearity criterion function $\Phi(\mathbf{w})$ (38) [7]. The exit criterion allows to determine the exit edge $\mathbf{r}_{k+1}$ (49) of the greatest descent of the collinearity criterion function $\Phi(\mathbf{w})$ (38). As a result of replacing the unit vector $\mathbf{e}_{i(k+1)}$ with the feature vector $\mathbf{x}_{k+1,}$ the value $\Phi(\mathbf{w}_k)$ of the collinearity function $\Phi(\mathbf{w})$ (38) decreases (41):

$$\Phi(\mathbf{w}_{k+1}) \leq \Phi(\mathbf{w}_k) - \varphi_{k+1}(\mathbf{w}_k) \tag{62}$$

After a finite number L ($L \leq m$) of the steps k, the collinearity function $\Phi(\mathbf{w})$ (38) reaches its minimum (61) at the final vertex $\mathbf{w}_L$ (59).

The sequence (60) of the inverse matrices $\boldsymbol{B}_k^{-1}$ is obtained in a multi-step process of minimizing the function $\Phi(\mathbf{w})$ (38). During the k-th step, the matrix $\boldsymbol{B}_{k-1} = [\mathbf{x}_1, \ldots, \mathbf{x}_{k-1}, \mathbf{e}_{i(k)}, \ldots, \mathbf{e}_{i(n)}]^T$ (12) is transformed into the matrix $\boldsymbol{B}_k$ by replacing the unit vector $\mathbf{e}_{i(k)}$ with the feature vector $\mathbf{x}_{k:}$

$$(\forall k \in \{1, \ldots, L\})\ \boldsymbol{B}_{k-1} \rightarrow \boldsymbol{B}_k \tag{63}$$

According to the vector Gauss-Jordan transformation, replacing the unit vector $\mathbf{e}_{i(k)}$ with the feature vector $\mathbf{x}_k$ during the k - th stage results in the following modifications of the co + lumns $\mathbf{r}_i(k)$ of the inverse matrix $\boldsymbol{B}_l^{-1} = [\mathbf{x}_1, \ldots, \mathbf{x}_l, \mathbf{e}_{i(l+1)}, \ldots, \mathbf{e}_{i(n)}]^T$ (49) [6]:

$$\mathbf{r}_{i(l+1)}(l+1) = \left(1/\mathbf{r}_{i(l+1)}(l)^T\mathbf{x}_{l+1}\right)\ \mathbf{r}_{i(l+1)}(l) \tag{64}$$

and

$$(\forall i \neq i(l+1))\ \mathbf{r}_i(l+1) = \mathbf{r}_i(l) - \left(\mathbf{r}_i(l)^T\mathbf{x}_{l+1}\right)\mathbf{r}_{i(l)}(l+1) =$$
$$= \mathbf{r}_i(l) - \left(\mathbf{r}_i(l)^T\mathbf{x}_{j(l+1)} / \mathbf{r}_{i(l)}(l)^T\mathbf{x}_{l+1}\right)\mathbf{r}_{i(l)}(l)$$

where $i(l+1)$ is the index of the unit vector $\mathbf{e}_{i(l+1)}$ leaving the basis $\boldsymbol{B}_l = [\mathbf{x}_1,\dots, \mathbf{x}_l, \mathbf{e}_{i(l+1)},\dots, \mathbf{e}_{i(n)}]^T$ during the l-th stage.

Remark 2: The vector Gauss-Jordan transformation (64) resulting from the replacing of the unit vector $\mathbf{e}_{i(k)}$ with the feature vector $\mathbf{x}_k$ in the basis $\boldsymbol{B}_{k-1} = [\mathbf{x}_1,\dots, \mathbf{x}_{k-1}, \mathbf{e}_{i(k)},\dots, \mathbf{e}_{i(n)}]^T$ cannot be executed when the below *collinearity condition* is met [7]:

$$\mathbf{r}_{i(k)}(k)^T\,\mathbf{x}_k = 0 \tag{65}$$

The collinearity condition (65) causes a division by zero in Eq. (64).

Let the symbol $\mathbf{r}_l[k]$ denote the l-th column $\mathbf{r}_l(k) = [r_{l,1}(k),\dots, r_{l,n}(k)]^T$ of the inverse matrix $\boldsymbol{B}_k^{-1} = [\mathbf{r}_1(k), \dots, \mathbf{r}_{k-1}(k), \mathbf{r}_k(k), \dots, \mathbf{r}_n(k)]$ (49) after the reduction of the last $n - k$ components $r_{l,i}(k)$:

$$\mathbf{r}_l[k] = [r_{l,1}(k), \dots, r_{l,k}(k)]^T \tag{66}$$

Similarly, the symbol $\mathbf{x}_j[k] = [x_{j,1},\dots,x_{j,k}]^T$ means the reduced vector obtained from the feature vector $\mathbf{x}_j = [x_{j,1},\dots,x_{j,n}]^T$ after he reducing of the last $n - k$ components $x_{j,i}$:

$$(\forall j \in \{1, \dots, m)\ \mathbf{x}_j[k] = \left[x_{j,1}, \dots, x_{j,k}\right]^T \tag{67}$$

Lemma 3: The collinearity condition (65) appears during the k-th step when the reduced vector $\mathbf{x}_k[k]$ (66) is a linear combination of the basis reduced vectors $\mathbf{x}_j[k]$ (67) with $j < k$:

$$\mathbf{x}_k[k] = \alpha_1\,\mathbf{x}_1[k] + \dots + \alpha_{k-1}\,\mathbf{x}_{k-1}[k] \tag{68}$$

where $(\forall i \in \{1, \dots, k - 1\})\ \alpha_i \in R^1$.

The proof of this lemma results directly from the collinearity condition (65) [7].

7. Small samples of multivariate feature vectors

A small sample of multivariate vectors appears when the number m of feature vectors $\mathbf{x}_j$ in the data set $\boldsymbol{C}$ (1) is much smaller than the dimension n of these vectors ($m << n$). The basis exchange algorithms allows for efficient minimization of the *CPL* criterion functions also in the case of small samples of multivariate vectors [10]. However, for small samples, some new properties of the basis exchange algorithms are more important. In particular, the regularization (42) of the *CPL* criterion functions becomes crucial. New properties of the basis exchange algorithms in the case of a small number m of multidimensional feature vectors $\mathbf{x}_j$ (1) is discussed on the example of the collinearity criterion function $\Phi(\mathbf{w})$ (38) and the regularized criterion function $\Psi(\mathbf{w})$ (42).

Lemma 4: The value $\Phi(\mathbf{w}_K)$ of the collinearity criterion function $\Phi(\mathbf{w})$ (38) at the final vertex $\mathbf{w}_L$ (59) is equal to zero if all m linear Eqs. (45) are fulfilled in the vertex

$\mathbf{w}_L$ which is related by the Eq. (47) to the matrix $\boldsymbol{B}_L = [\mathbf{x}_1,..., \mathbf{x}_m, \mathbf{e}_{i(1)},..., \mathbf{e}_{i(n-m)}]^T$ (48) containing the unit vectors $\mathbf{e}_i$ with the indices i from the subset I_L $(i \in I_L)$.

Theorem 4: If the feature vectors $\mathbf{x}_j$ constituting the subset $\boldsymbol{C}_k$ ($\boldsymbol{C}_k \subset \boldsymbol{C}$ (1)) and used in the definition of the function $\Phi(\mathbf{w})$ (38) are linearly independent (*Def.* 2), then the value $\Phi(\mathbf{w}_L)$ of the collinearity criterion function $\Phi(\mathbf{w})$ at the final vertex $\mathbf{w}_L$ (59) is equal to zero ($\Phi(\mathbf{w}_L) = 0$).

The proof of Theorem 4 can be based on the stepwise inversion of the matrices $\boldsymbol{B}_k$ (48) [16]. The final vertex $\mathbf{w}_L$ (59) can be found by inverting the related matrix $\boldsymbol{B}_L = [\mathbf{x}_1,..., \mathbf{x}_{rk}, \mathbf{e}_{i(1)},..., \mathbf{e}_{i(n-rk)}]^T$ (48).

The final vertex $\mathbf{w}_L$ (59) resetting ($\Phi(\mathbf{w}_L) = 0$) the criterion function $\Phi(\mathbf{w})$ (38) can be related to the optimal matrix $\boldsymbol{B}_L = [\mathbf{x}_1,..., \mathbf{x}_L, \mathrm{e}_{i(L+1)},..., \mathrm{e}_{i(n)}]^T$ (48) built from L ($L \leq m$) feature vectors $\mathbf{x}_j$ ($j \in J(\mathbf{w}_L)$ (45)) from the data set $\boldsymbol{C}$ (1) and from n - L selected unit vectors $\mathbf{e}_i$ ($i \in I(\mathbf{w}_L)$ (46)). Different subsets of the unit vectors $\mathbf{e}_i$ in the final matrix $\boldsymbol{B}_L$ (48) result in different positions of the final vertices $\mathbf{w}_{L(l)}$ (59) in the parameter space $\boldsymbol{R}^n$. The criterion function $\Phi(\mathbf{w})$ (38) is equal zero ($\Phi(\mathbf{w}_{L(l)}) = 0$) at each of these vertices $\mathbf{w}_{L(l)}$ (59).

The position of the final vertices $\mathbf{w}_{L(l)}$ (59) in the parameter space $\boldsymbol{R}^n$ depends on which unit vectors $\mathbf{e}_i$ ($i \in I_{L(l)}$) are included in the basis $\mathbf{B}_{L(l)}$ (48), where:

$$(\forall l \in \{1, \dots, l_{\max}\})\ \Phi_k(\mathbf{w}_{L(l)}) = 0 \tag{69}$$

The maximal number $l_{\max}$ (69) of different vertices $\mathbf{w}_{L(l)}$ (59) can be large when $m << n$:

$$l_{\max} = n!/m!(n-m)! \tag{70}$$

The choice between different final vertices $\mathbf{w}_{L(l)}$ (59) can be based on the minimization of the regularized criterion function $\Psi(\mathbf{w})$ (42). The regularized function $\Psi(\mathbf{w})$ (42) is the sum of the collinearity function $\Phi(\mathbf{w})$ (38) and the weighted sum of the cost functions $\varphi_i^0(\mathbf{w})$ (43). If $\Phi(\mathbf{w}_{L(l)}) = 0$ (38), then the value $\Psi(\mathbf{w}_{L(l)})$ of the criterion function $\Psi(\mathbf{w})$ (42) at the final vertex $\mathbf{w}_{L(l)}$ (59) can be given as follows:

$$\begin{aligned}\Psi(\mathbf{w}_{L(l)}) &= \lambda_i\, \Sigma_i\, \chi_i\, \varphi_j^{\,0}(\mathbf{w}_{L(l)}) = \\ &= \lambda\, \Sigma\, \chi_i \mid \mathrm{w}_{L(l),i} \mid\end{aligned} \tag{71}$$

where the above sums take into account only the indices i of the subset $I(\mathbf{w}_{L(l)})$ of the non-zero components $\mathrm{w}_{L(l),i}$ of the final vertex $\mathbf{w}_{L(l)} = [\mathrm{w}_{L(l),1}, \dots, \mathrm{w}_{L(l),n}]^T$ (59):

$$I(\mathbf{w}_{L(l)}) = \{i : \mathbf{e}_i^{\,T}\mathbf{w}_{L(l)} \neq 0\} = \{i : \mathrm{w}_{L(l),i} \neq 0\} \tag{72}$$

If the final vertex $\mathbf{w}_{L(l)}$ (59) is not degenerate (*Def.* 5), then the matrix $\boldsymbol{B}_{L(l)}$ (48) is built from all m feature vectors $\mathbf{x}_j$ ($j \in \{1,...., m\}$) making up the data set $\boldsymbol{C}$ (1) and from n - m selected unit vectors $\mathbf{e}_i$ ($i \in I(\mathbf{w}_{L(l)})$ (71)).

$$\boldsymbol{B}_{)m} = [\mathbf{x}_1, \dots, \mathbf{x}_m, \mathbf{e}_{i(m+1)}, \dots, \mathbf{e}_{i(n)}]^T \tag{73}$$

The problem of the constrained minimizing of the regularized function $\Psi(\mathbf{w})$ (71) at the vertices $\mathbf{w}_{L(l)}$ (59) satisfying the conditions $\Phi(\mathbf{w}_{L(l)}) = 0$ (69) can be formulated in the following way:

$$\begin{aligned}&min_l\{\Psi(\mathbf{w}_{L(l)}): \Phi(\mathbf{w}_{L(l)}) = 0\} = \\ &= min_l\{\Sigma_i\, \gamma_i \mid \mathrm{w}_{L(l),i} \mid: \Phi(\mathbf{w}_{L(l)}) = 0\}\end{aligned} \tag{74}$$

According to the above formulation, the search for the minimum of the regularized criterion function $\Psi(\mathbf{w})$ (42) is takes place at all such vertices $\mathbf{w}_{L(l)}$ (59), where the collinearity function $\Phi(\mathbf{w})$ (38) is equal to zero. The regularized criterion function $\Psi(\mathbf{w})$ (42) is defined as follows at the final vertices $\mathbf{w}_{L(l)} = [w_{L(l),1}, \ldots, w_{L(l),n}]^T$ (59), where $\Phi(\mathbf{w}_{L(l)}) = 0$:

$$(\forall \mathbf{w}_{L(l)})\ \Psi'(\mathbf{w}_{L(l)}) = \Sigma\, \gamma_i \mid w_{L(l),i} \mid \tag{75}$$

The optimal vertex $\mathbf{w}_{L(l)}^*$ is the minimum value $\Psi'(\mathbf{w}_{L(l)}^*)$ of the *CPL* criterion function $\Psi'(\mathbf{w})$ (75) defined on such final vertices $\mathbf{w}_{L(l)}$ (59), where $\Phi(\mathbf{w}_{L(l)}) = 0$ (38):

$$(\exists \mathbf{w}_{L(l)}^*)\ (\forall \mathbf{w}_{L(l)} : \Phi(\mathbf{w}_{L(l)}) = 0)\ \Psi'(\mathbf{w}_{L(l)}) \geq \Psi'(\mathbf{w}_{L(l)}^*) > 0 \tag{76}$$

As in the case of the minimization of the perceptron criterion function $\Phi_k^P(\mathbf{v})$ (29), the optimal vector $\mathbf{w}_{L(l)}^*$ (76) may be located at a selected vertex of some convex polyhedron (27) in the parameter space $\boldsymbol{R}^n$ ($\mathbf{w} \in \boldsymbol{R}^n$) [7].

If the cost parameters γ_i (42) have standard values of one ($(\forall i \in \{1, \ldots, n\})\ \gamma_i = 1$), then the constraint minimization problem (74) leads to the optimal vertex $\mathbf{w}_{L(l)}^*$ with the smallest L_1 length $\| \mathbf{w}_{L(l)}^* \|_{L1} = |w_{L(l),1}^*| + \ldots + |w_{L(l),n}^*|$, where $\Phi(\mathbf{w}_{L(l)}^*) = 0$ (38):

$$(\exists \mathbf{w}_{L(l)}^*)\ (\forall \mathbf{w}_{L(l)} : \Phi(\mathbf{w}_{L(l)}) = 0)\ \|\mathbf{w}_{L(l)}\| \geq \|\mathbf{w}_{L(l)}^*\| \tag{77}$$

Optimal vertex $\mathbf{w}_{L(l)}^*$ with the smallest L_1 length $\| \mathbf{w}_{L(l)}^* \|_{L1}$ (77) is related to the largest L_1 margin $\delta_{L1}(\mathbf{w}_{L(l)}^*)$ (6) [11]:

$$\delta_{L1}(\mathbf{w}_{L(l)}^*) = 2/\| \mathbf{w}_{L(l)}^* \|_{L1} = 2/(|w_{L(l),1}^*| + \ldots + |w_{L(l),n}^*|) \tag{78}$$

The basis exchane algorithm allow to solve the constraint minimization problem (74) and to find the optimal vertex $\mathbf{w}_{L(l)}^*$ (77) with the largest L_1 margin $\delta_{L1}(\mathbf{w}_{L(l)}^*)$.

Support Vector Machines (*SVM*) is the most popular method for designing linear classifiers or prognostic models with large margins [12]. According to the *SVM* approach, the optimal linear classifier or the prognostic model defined by such an optimal weight vector $\mathbf{w}^*$ that has a maximum margin $\delta_{L2}(\mathbf{w}^*)$ based on the Euclidean (L_2) norm:

$$\delta_{L2}(\mathbf{w}^*) = 2/\| \mathbf{w}^* \|_{L2} = 2/\left((\mathbf{w}^*)^T \mathbf{w}^*\right)^{1/2} \tag{79}$$

Maximization of the Euclidean margins $\delta_{L2}(\mathbf{w})$ (79) is performed using quadratic programming [2].

8. Complex layers of linear prognostic models

Complex layers of linear classifiers or prognostic models have been proposed as a scheme for obtaining a general classification or forecasting rules designed on the basis of a small number of multidimensional feature vectors $\mathbf{x}_j$ [11]. According to this scheme, when designing linear prognostic models, averaging over a small number m of feature vectors $\mathbf{x}_j$ of the dimension n ($m << n$) is replaced by averaging on collinear clusters of selected features (genes) X_i. Such an approach to averaging can be linked to the ergodic theory [17].

In the case of a small sample of multivariate vectors, the number m of feature vectors $\mathbf{x}_j$ in the data set $\boldsymbol{C}$ (1) may be much smaller than the dimension n of these vectors ($m << n$). In this case, the collinear cluster $C(\mathbf{w}_k^*[r_k])$ (52) may contain all feature vectors $\mathbf{x}_j$ from the set C (1) and the vertex $\mathbf{w}_k^*[r_k]$ may have the rank r_k equal to m ($r_k = m$).

As it follows from Theorem 4, if the collinearity criterion function $\Phi(\mathbf{w})$ (38) is defined on linearly independent (*Def.* 2) feature vectors $\mathbf{x}_j$, then the values $\Phi(\mathbf{w}_{m(l)})$ of this function at each final vertex $\mathbf{w}_{m(l)}$ (59) are equal to zero ($\Phi(\mathbf{w}_{m(l)}) = 0$). Each final vertex $\mathbf{w}_{m(l)}$ (59) can be reached in m steps k ($k = 1, \ldots, m$) starting from the vertex $\mathbf{w}_0 = [0,\ldots, 0]^T$ related to the identity matrix $B_0 = I_n = [\mathbf{e}_1,\ldots, \mathbf{e}_n]^T$.

Minimization of the collinearity criterion function $\Phi(\mathbf{w})$ (38), and then minimization of the criterion function $\Psi'(\mathbf{w}_{L(l)})$ (75) at the final vertices $\mathbf{w}_{L(l)}$ (59) allows to determine the optimal vertex $\mathbf{w}_{L(l)}^*$ (77) with the largest L_1 margin $\delta_{L1}(\mathbf{w}_{L(l)}^*)$ (78). If the feature vectors $\mathbf{x}_j$ (1) are linearly independent, then the optimal vertex $\mathbf{w}_{L(l)}^*$ (77) is related to the optimal basis $B_{L(l)}^* = [\mathbf{x}_1,\ldots, \mathbf{x}_m, \mathbf{e}_{i(m+1)},\ldots, \mathbf{e}_{i(n)}]^T$ which contains all m feature vectors $\mathbf{x}_j$ (1) and $n - m$ unit vectors $\mathbf{e}_i$ with the indices i belonging to the optimal subset $I(\mathbf{w}_{L(l)}^*)$ (71) ($i \in I(\mathbf{w}_{L(l)}^*)$).

The optimal basis $B_m^* = [\mathbf{x}_1,\ldots, \mathbf{x}_m, \mathbf{e}_{i(m+1)},\ldots, \mathbf{e}_{i(n)}]^T$ (73) is found in two stages. In the first stage, m feature vectors $\mathbf{x}_j$ (1) are introduced into matrices $B_k = [\mathbf{x}_1,\ldots, \mathbf{x}_k, \mathbf{e}_{i(k+1)},\ldots, \mathbf{e}_{i(n)}]^T$ ($k = 0, 1, \ldots, m - 1$). The inverse matrices B_k^{-1} (49) are computed in accordance with the vector Gauss-Jordan transformation (64). In the second stage, the unit vectors $\mathbf{e}_{i(l)}$ in the matrices $B_{m(l)}$ (73) are exchanged to minimize the *CPL* function $\Psi'(\mathbf{w}_{m(l)})$ (75) at the final vertices $\mathbf{w}_{m(l)}$ (77). The optimal basis B_m^* defines (47) the optimal vertex $\mathbf{w}_{m(l)}^*$ (77), which is characterized by the largest margin $\delta_{L1}(\mathbf{w}_{m(l)}^*)$ (78).

The vertexical feature subspace $F_1^*[m]$ ($F_1^*[m] \subset F[n]$ (1)) can be obtained on the basis of the optimal vertex $\mathbf{w}_{m(l)}^*$ (77) with the largest margin $\delta_{L1}(\mathbf{w}_{m(l)}^*)$ (78). The vertexical subspace $F_1^*[m]$ contains the reduced vectors $\mathbf{x}_{1,j}[m]$ with the dimension m [7]:

$$(\forall j \in \{1, \ldots, m\})\ \mathbf{x}_{1,j}[m] \in F_1^*[m] \tag{80}$$

The reduced vectors $\mathbf{x}_{1,j}[m]$ (80) are obtained from the feature vectors $\mathbf{x}_j = [x_{j,1},\ldots,x_{j,n}]^T$ ($\mathbf{x}_j \in F[n]$) ignoring such components $x_{j,i}$ which are related to the unit vectors $\mathbf{e}_i$ in the optimal basis $B_1^* = [\mathbf{x}_1,\ldots, \mathbf{x}_m, \mathbf{e}_{i(m+1)},\ldots, \mathbf{e}_{i(n)}]^T$ (73). The reduced vectors $\mathbf{x}_{1,j}[m]$ are represented by such m features X_i ($X_i \in R_1^*$ (54)), which are not linked to the unit vectors $\mathbf{e}_i$ ($i \notin I_{m(l)}^*$) in the basis $B_{m(l)}^*$ (73) representing the optimal vertex $\mathbf{w}_{m(l)}^*$ (77).

$$R_1^* = \{X_{i(1)}, \ldots, X_{i(m)} : i(l) \notin I_{m(l)}^*\ (72)\} \tag{81}$$

The m features $X_{i(l)}$ belonging to the optimal subset R_1^* ($X_{i(l)} \in R_1^*$ (81) are related to the weights $w_{k.l}^*$ ($\mathbf{w}_k^*[m] = [w_{k,1}^*, \ldots, w_{k,m}^*]^T$) that are not zero ($w_{k.l}^* \neq 0$).

The optimal feature subset R_1^* (81) consists of m collinear features X_i. The optimal vertex $\mathbf{w}_1^*[m]$ ($\Phi(\mathbf{w}_1^*[m]) = 0$ (69)) in the reduced parameter space R^m ($\mathbf{w}_1^*[m] \in R^m$) is based on these m features X_i. The reduced optimal vertex $\mathbf{w}_1^*[m]$ with the largest margin $\delta_{L1}(\mathbf{w}_1^*[m])$ (77) is the unique solution of the constrained optimization problem (74). Maximizing the L_1 margin $\delta_{L1}(\mathbf{w}_l^*)$ (78) leads to the first reduced vertex $\mathbf{w}_1^*[m] = [w_{k,1}^*, \ldots, w_{k,m}^*]^T$ with non-zero components $w_{k.i}^*$ ($w_{k.i}^* \neq 0$).

The collinear interaction model between m collinear features $X_{i(l)}$ from the optimal subset $R_1^*(m)$ (81) can be formulated as follows (57):

$$w_{k.1}^*\, X_{i(1)} + \ldots + w_{k.m}^*\, X_{i(m)} = 1 \tag{82}$$

The prognostic models for each feature $X_{i'}$ from the subset R_1^* (81) may have the following form (58):

$$(\forall i' \in \{1, \ldots, m) \\ X_{i'} = \alpha_{i',0} + \alpha_{i',1} X_{i(1)} + \ldots + \alpha_{i',m} X_{i(m)} \tag{83}$$

where $\alpha_{i',0} = 1 / w_{k.i'}^*$, $\alpha_{i',i'} = 0$, and $(\forall\, i(l) \neq i')\ \alpha_{i',i(l)} = w_{k.i(l)}^* / w_{k.i'}^*$.

In the case of a data set **C** with a small number m ($m << n$) of multidimensional feature vectors $\mathbf{x}_j$ (1), the prognostic models (83) for individual features $X_{i'}$ can be weak. It is know that sets (ensembles) of weak models can have strong generalizing properties [4]. A set of weak prognostic models (83) for a selected feature (dependent variable) $X_{i'}$ can be implemented in the complex layer of L prognostic models (83) [11].

The complex layer can be built on the basis of the sequence of L optimal vertices $\mathbf{w}_l^*$ (77) related to m features X_i constituting the subsets R_l^* (81), where $l = 0, 1, \ldots, L$.

$$(\mathbf{w}_1^*, R_1^*), \ldots, (\mathbf{w}_L^*, R_L^*) \tag{84}$$

Design assumption: Each subset R_l^* (81) in the sequence (84) contains a priori selected feature (dependent variable) $X_{i'}$ and $m - 1$ other features (independent variables) $X_{i(l)}$. The other features $X_{i(l)}$ ($X_{i(l)} \in R_l^*$) should be different in successive subsets R_l^* ($l = 0, 1, \ldots, L$).

The first optimal; vertex $\mathbf{w}_1^*$ (77) in the sequence (84) is designed on the basis of m feature vectors $\mathbf{x}_j$ (1), which are represented by all n features X_i constituting the feature set $F(n) = \{X_1, \ldots, X_n\}$. The vertex $\mathbf{w}_1^*$ (77) is found by solving the constraint optimization problem (74) according to the procedure with the two stages outlined earlier. The two-stage procedure allows to find the optimal vertex $\mathbf{w}_1^*$ (77) with the largest L_1 margin $\delta_{L1}(\mathbf{w}_1^*)$ (78).

The second optimal vertex $\mathbf{w}_2^*$ (77) in the sequence (84) is obtained on the basis of m reduced feature vectors $\mathbf{x}_j[n - (m - 1)]$ (67), which are represented by $n - (m - 1)$ features X_i constituting the reduced feature subset $F_2(n - (m + 1))$:

$$F_2(n - (m - 1)) = F(n)/R_1^* \cup \{X_{i'}\} \tag{85}$$

The l-th optimal vertex $\mathbf{w}_l^*$ (77) in the sequence (84) is designed on the basis of m reduced vectors $\mathbf{x}_j[n - l(m - 1)]$ (67), which are represented by $n - l(m - 1)$ features X_i constituting the feature subset $F_l(n - l(m - 1))$:

$$F_l(n - l(m - 1)) = F_{l-1}(n - l(m - 1))/R_{l-1}^* \cup \{X_{i'}\} \tag{86}$$

The sequence (84) of L optimal vertices $\mathbf{w}_l^*$ (77) related to the subsets $F_l(n - l(m - 1))$ (86) of features is characterized by decreased L_1 margins $\delta_{L1}(\mathbf{w}_l^*)$ (78) [18].

$$\delta_{L1}(\mathbf{w}_1^*) \geq \delta_{L1}(\mathbf{w}_2^*) \geq \ldots \geq \delta_{L1}(\mathbf{w}_L^*) \tag{87}$$

The prognostic models (83) for the dependent feature (variable) $X_{i'}$ are designed for each subset $F_l(n - l(m - 1))$ (86) of features X_i, where $l = 0, 1, \ldots, L$ (84):

$$(\forall l \in \{0, 1, \ldots, L\} \\ X_{i'}(l) = \alpha_{i',0}(l) + \alpha_{i',1}(l)\, X_{i(1)}(l) + \ldots + \alpha_{i',m}(l)\, X_{i(m)} \tag{88}$$

The final forecast $X_{i'}^\wedge$ for the dependent feature (variable) $X_{i'}$ based on the complex layer of $L + 1$ prognostic models (88) can have the following form:

$$X_{i'}{}^{\wedge} = (X_{i'}(1) + \ldots + X_{i'}(L))/(L+1) \tag{89}$$

In accordance with the Eq. (89), the final forecast $X_{i(m)}{}^{\wedge}$ for the feature $X_{i'}$ results from averaging the forecasts of $L + 1$ individual models $X_{i'}(l)$ (88).

9. Concluding remarks

The article considers computational schemes of designing classifiers or prognostic models based on such a data set **C** (1), which consists of a small number m of high-dimensional feature vectors $\mathbf{x}_j$ ($m < < n$).

The concept of a complex layer composed of many linear prognostic models (88) built in low-dimensional feature subspaces is discussed in more detail. These models (88) are built by using a small number m of collinear features X_i belonging to the optimal feature clusters $R_l{}^*$ (81). The optimal feature clusters $R_l{}^*$ (81) are formed by the search for the largest margins $\delta_{L1}(\mathbf{w}_l{}^*)$ (78) in the L_1 norm.

The averaged prognostic models $X_{i'}{}^{\wedge}$ (89) are based on the layer of L parallel models $X_{i'}(l)$ (88). In line with the ergodic theory, averaging on a small number m of feature vectors $\mathbf{x}_j$ has been replaced with averaging on L collinear clusters $R_l{}^*$ (81) of features X_i. Such averaging scheme should allow for a more stable extraction of general patterns from small samples of high-dimensional feature vectors $\mathbf{x}_j$ (1) [11].

Author details

Leon Bobrowski[1,2]

1 Faculty of Computer Science, Białystok University of Technology, Poland

2 Institute of Biocybernetics and Biomedical Engineering, PAS, Warsaw, Poland

*Address all correspondence to: l.bobrowski@pb.edu.pl

References

[1] Duda O. R., Hart P. E., and Stork D. G., *Pattern classification*, J. Wiley, New York, 2001

[2] Hand D., Smyth P., and Mannila H., *Principles of data mining*, MIT Press, Cambridge (2001)

[3] Bishop C. M., *Pattern Recognition and Machine Learning*. Springer Verlag, 2006

[4] Kuncheva L.: *Combining Pattern Classifiers: Methods and Algorithms*, 2nd Edition, J. Wiley, New Jersey (2014).

[5] Simonnard M., *Linear Programming*, Prentice – Hall, New York, Englewood Cliffs, 1966

[6] Bobrowski L., *Data mining based on convex and piecewise linear* (CPL) *criterion functions* (*in Polish*), Białystok University of Technology, 2005

[7] Bobrowski L., *Data Exploration and Linear Separability*, pp. 1 - 172, Lambert Academic Publishing, 2019

[8] Bobrowski, L.: "Design of piecewise linear classifiers from formal neurons by some basis exchange technique", *Pattern Recognition*, 24(9), pp. 863-870 (1991).

[9] Bobrowski L., Zabielski P., "Models of Multiple Interactions from Collinear Patterns", pp. 153-165 in: *Bioinformatics and Biomedical Engineering* (*IWBBIO* 2018), Eds.: I. Rojas, F. Guzman, LNCS 10208, Springer Verlag, 2018

[10] Bobrowski L., Small Samples of Multidimensional Feature Vectors (*ICCCI 2020*), pp. 87 - 98 in: *Advances in Computational Collective Intelligence*, Eds.: Hernes M, et al., Springer 2020

[11] Bobrowski L., "Complexes of Low Dimensional Linear Classifiers with L_1 Margins", pp. *29 - 40 in:* ACIIDS 2021, Springer Verlag, 2021

[12] Boser B. E., Guyon I., Vapnik V. N.: A training algorithm for optimal margin classifiers. In: Proceedings of the Fifth *Annual Workshop of Computational Learning Theory*, 5, 144–152. Pittsburgh, ACM, 1992

[13] Bobrowski L., Łukaszuk T.: Repeatable functionalities in complex layers of formal neurons, *EANN* 2021, *Engineering Applications of Neural Networks*, Springer 2021

[14] Rosenblatt F.: *Principles of neurodynamics*, Spartan Books, Washington, 1962

[15] Bobrowski L., Łukaszuk, T.: Relaxed Linear Separability (*RLS*) Approach to Feature (Gene) Subset Selection, pp. 103 - 118 in: *Selected Works in Bioinformatics*, Edited by: Xuhua Xia, *INTECH*, 2011

[16] Bobrowski L.: "Large Matrices Inversion Using the Basis Exchange Algorithm", *British Journal of Mathematics & Computer Science*, 21(1): 1-11, 2017

[17] Petersen K.: *Ergodic Theory* (*Cambridge Studies in Advanced Mathematics*), Cambridge University Press, 1990

[18] Bobrowski L., Zabielski P.: "Feature (gene) clustering with collinearity models", ***ICCCI* 2021** (*to appear*), Springer Verlag, 2021

Section 2

Applications of Data Mining

Chapter 8

Artificial Intelligence and Its Application in Optimization under Uncertainty

Saeid Sadeghi, Maghsoud Amiri and Farzaneh Mansoori Mooseloo

Abstract

Nowadays, the increase in data acquisition and availability and complexity around optimization make it imperative to jointly use artificial intelligence (AI) and optimization for devising data-driven and intelligent decision support systems (DSS). A DSS can be successful if large amounts of interactive data proceed fast and robustly and extract useful information and knowledge to help decision-making. In this context, the data-driven approach has gained prominence due to its provision of insights for decision-making and easy implementation. The data-driven approach can discover various database patterns without relying on prior knowledge while also handling flexible objectives and multiple scenarios. This chapter reviews recent advances in data-driven optimization, highlighting the promise of data-driven optimization that integrates mathematical programming and machine learning (ML) for decision-making under uncertainty and identifies potential research opportunities. This chapter provides guidelines and implications for researchers, managers, and practitioners in operations research who want to advance their decision-making capabilities under uncertainty concerning data-driven optimization. Then, a comprehensive review and classification of the relevant publications on the data-driven stochastic program, data-driven robust optimization, and data-driven chance-constrained are presented. This chapter also identifies fertile avenues for future research that focus on deep-data-driven optimization, deep data-driven models, as well as online learning-based data-driven optimization. Perspectives on reinforcement learning (RL)-based data-driven optimization and deep RL for solving NP-hard problems are discussed. We investigate the application of data-driven optimization in different case studies to demonstrate improvements in operational performance over conventional optimization methodology. Finally, some managerial implications and some future directions are provided.

Keywords: Data-driven optimization, Decision making under uncertainty, Mathematical optimization, Machine learning, Deep learning, Reinforcement learning

1. Introduction

Optimization is applied in many engineering and science fields, including manufacturing, inventory control, transportation, finance, economics [1, 2].

Some parameters involved in optimization problems are subject to uncertainty in real practice due to various reasons, including measurement errors and uncontrollable disturbances [3]. Such uncertain parameters can be product demand and price, raw material supply chain cost, production cost. Disregarding uncertainty could, unfortunately, render the solution of a deterministic optimization problem suboptimal or even infeasible. In the era of big data and deep learning (DL), intelligent use of data and knowledge extraction from them have great benefits for organizations. Besides, in today's complex world, uncertainty on the lack of enough data has been replaced by too much data, which creates numerous opportunities for academicians and practitioners [4]. A large amount of interactive data is routinely created, collected, and archived in different industries; these data are becoming an important asset in process operation, control, and design. Explosive growth in volume and different sorts of data in organizations has created the need to develop technologies that can intelligently and rapidly analyze large volumes of data [4]. The traditional optimization methods cannot face big data satisfactorily. Nowadays, a wide array of emerging machine learning (ML) techniques can be leveraged to analyze data and extract relevant, accurate, and useful information and knowledge for smart decision-making. More recently, the dramatic progress of ML, especially DL over the past decade, coupled with recent advances in mathematical programming, sparks a flurry of interest in data-driven optimization [5, 6]. The uncertainty model is formulated based on a data-driven optimization paradigm, allowing uncertainty data to speak for themselves in the optimization algorithm. In this way, rich knowledge underlying uncertainty data set can be extracted and harnessed automatically for smart and data-driven decision making. In such situations, the effectiveness and efficiency of traditional operational research methods are questionable. In recent years, the inefficiency of traditional methods in facing the uncertainty caused by big data has led researchers to integrate *artificial intelligence* (AI) with optimization methods. Integrating AI and optimization methods play a crucial role in solving problems in dynamic and uncertain environments. Nowadays, a wide range of ML tools has emerged that can be leveraged to analyze data automatically and extract relevant, accurate, and useful information for smart and data-driven decision-making. DL is one of the most rapidly growing sub-fields of the ML technique that demonstrates remarkable power in processing and deciphering a large volume of data through a complex architecture. Reinforcement learning (RL) is another ML sub-field that recently is applied to tackle complex sequential decision problems. This branch of ML epitomizes a step toward building autonomous systems by understanding the visual world.

The objective of this study is to provide an overview of the use of data-driven optimization in academia and practice from the following perspectives:

1. How can integrate artificial intelligence techniques with mathematical programming models to develop the intelligencete and data-driven *decision support systems (DSS)* in uncertain conditions caused by big data?

2. We demonstrate the use of data-driven optimization across three case studies from operations research.

In this regard, this chapter reviews recent advances in data-driven optimization that highlight the integration of mathematical programming and ML for decision-making under uncertainty and identifies potential research opportunities. We compare data-driven optimization performance to conventional models from optimization methodology. We summarize the existing research papers on data-driven optimization under uncertainty and classify them into three categories: Data-driven stochastic program, Data-driven robust optimization, and Data-driven chance-constrained, according to their unique approach to uncertainty modeling distinct optimization structures. Based on the literature survey, we identify five promising future research directions on optimization under uncertainty in the era of big data and DL, (i) Employment of DL in the field of data-driven optimization under uncertainty, (ii) Deep data-driven models, (iii) Online learning-based data-driven optimization, (iv) Leveraging RL techniques for optimization, and (v) Deep RL for solving NP-hard problems and highlight respective research challenges and potential methodologies. We conducted an extensive literature review on recent papers published across the premier journals between 2002 and 2020 in our field, namely, the European Journal of Operational Research, Operations Research, Journal of Cleaner Production, Production and Operations Management, Journal of Operations Management, Computers in Industry, and Decision Sciences. We specifically searched for papers containing "big data", "data-driven optimization", "artificial intelligence", "machine learning", "deep learning", and "Reinforcement learning". However, our research into the existing literature reveals a scarcity of research works utilizing DL and RL in these disciplines.

The remainder of this paper is organized as follows: Section 2 provides an introduction to the mathematical optimization method. In Section 3, a brief review of AI methods such as ML, DL, and RL is provided. In sections 4–6, applying different ML, DL, and RL techniques in data-driven optimization under uncertainty are presented. Finally, the book chapter ends with the conclusion, some managerial implications, and future research recommendations.

2. Mathematical optimization under uncertainty

In recent years, mathematical programming techniques for decision-making under uncertainty have been applied in many science and engineering areas, including process design, production scheduling and planning, design, control, and supply chain optimization.

Optimization under uncertainty has been motivated because parameters involved in optimization models for design, planning, scheduling, and supply chains are often uncertain parameters such as product demands, prices of raw material, product, and yields.

A major modeling decision in optimization under uncertainty is whether the decision-maker should rely on robust optimization to use stochastic programming [7]. The robust optimization basis idea is to guarantee feasibility over a specified uncertainty set. In contrast, in the stochastic programming approach, a subset of decisions is set by anticipating that recourse actions can be taken once the uncertainties are revealed over a pre-specified scenario with discrete probabilities of uncertainties. The robust optimization basis idea is to guarantee feasibility over a specified uncertainty set. In contrast, in the stochastic programming approach, a subset of decisions is set by anticipating that recourse actions can be taken once the uncertainties are revealed over a pre-specified scenario with discrete probabilities of the uncertainties.

In general, the optimization approach tends to be more appropriate for short-term scheduling problems in which feasibility over a specified set of uncertain parameters is a major concern and when there is not much scope for recourse decisions. On the other hand, the stochastic programming approach tends to be more appropriate for long-term production planning and strategic design decisions.

In this section, the authors briefly explain three leading modeling paradigms for optimization under uncertainty, namely stochastic programming, robust optimization, and chance-constrained programming.

2.1 Stochastic programming

Under uncertainty, a common decision-making approach is stochastic programming, aiming to optimize the expected objective value across all the uncertainty realizations [8]. The stochastic programming key idea is to model the randomness in uncertain parameters with probability distributions. In this approach, the first stage, all the decisions must be made without knowing precisely the uncertainty realizations. The decision-maker then waits for resolving the uncertainty and knowing the actual value of the uncertain parameters. In the second stage, the decision-maker takes corrective actions after uncertainty is revealed. The stochastic programming approach has demonstrated various applications, such as inventory routing problems [9], supply chain network modeling [10], distributed energy systems design [11], optimal tactical planning [12], and energy management [13].

2.2 Robust optimization

Robust optimization is a promising alternative paradigm to optimization under uncertainty that does not require accurate knowledge on probability distributions of uncertain parameters. The key idea of robust optimization is to construct a convex uncertainty set of possible realizations of the uncertain parameters and then optimize against worse case realization within this set [14]. A robust optimization framework aims to hedge against the worst-case within the uncertainty set. The robust optimization approach has demonstrated various applications, such as supply chain planning [15], supply chain management [16], inventory management [17].

2.3 Chance constrained programming

Chance constrained programming is another common paradigm for optimization under uncertainty with soft probabilistic constraints on the decision variable in place of the hard ones present in robust optimization. Specifically, chance-constrained programming aims to compute a solution that satisfies the constraint with high probability in an uncertain environment. In the chance-constrained optimization paradigm, the probability distribution of uncertain parameters should be known to capture the randomness of uncertain parameters. Chance constrained programs are increasingly used in many applications, such as robotics [18], stochastic model predictive control [19], energy systems [20], and autonomous driving [21].

All mathematical optimization methods are inefficient and effective in facing uncertainty caused by the large volume of data. In the following section, three AI areas as tools for compensating the weaknesses of mathematical optimizing methods are introduced. The term "AI" is often used to describe machines (or computers) that mimic "cognitive" functions that humans associate with the human mind, such as "learning" and problem-solving" [22]. A brief description of the three main areas of AI, including ML, DL, and RL, is provided in the following.

3. Machine learning (ML)

ML is a sub-area of AI that can automatically extract artificial information and knowledge from diverse data types with high speed. The advancement in computational power and the emergence of big data have led to ML optimization and simulation methods. Analysis of big data by ML offers considerable advantages for integrating and evaluating large amounts of complex data [23]. ML solutions have scalability and flexibility compared with traditional statistical methods, making them deployable for many tasks, such as clustering, classification, and prediction. ML models have demonstrated outstanding ability for learning intricate patterns that enable them to make predictions about unobserved data. In addition to using models for prediction, it can accurately interpret what a model has learned.

ML techniques use large sets of data inputs and outputs to recognize patterns and effectively "learn" to make autonomous recommendations or decisions [24]. These algorithms attempt to minimize their errors and maximize the likelihood of their predictions being true [25]. The predictive abilities of ML models are increasingly applied in various fields such as healthcare, genetic, finance, education, and production.

3.1 Deep learning (DL)

In real applications, uncertainty data exhibit highly complex and nonlinear characteristics. DL is an ML technique and includes algorithms and computational models that imitate the architecture of the biological neural networks in the brain [artificial neural networks (ANNs)] [25]. The DL technology consists of numerous layers responsible for extracting important abstract features from the data [26]. It can process a large volume of data through a complex architecture [27]. DL algorithms can uncover useful uncertainty data patterns for mathematical programming [28]. Recently, the DL technique has been used in optimization under uncertainty.

3.2 Reinforcement learning (RL)

In particular, RL has gained tremendous attraction recently in different research areas. In RL, an agent gains experience from directly interacting with the environment and selecting an optimal action. RL is concerned with how a software agent should choose an action to maximize a cumulative reward. Combining DL with the RL technique creates the concept of deep RL, which enables RL to tackle the previously intractable decision-making problems. Inspired by the recent advances of deep RL in video games, robotics, and cyber-security, it has been used in optimization problems.

After introducing mathematical optimization methods and three main AI areas, it is time to pay to apply ML, DL, and RL methods in data-driven optimization. They are discussed in turn in the following sections.

4. Leveraging ML techniques for hedging against uncertainty in data-driven optimization

In the big data and ML era, a large amount of interactive data are routinely generated and collected in different industries. Intelligence and data-driven analysis and decision-making have a critical role in process operations, design, and control. The success of the DSS depends primarily on the ability to process and analyze large amounts of data and extract relevant and useful knowledge and information from them. In this context, the data-driven approach has gained prominence due to its provision of insights for decision-making and easy implementation. The data-driven optimization framework is a hybrid system that integrates AI and optimization methods for devising a data-driven and intelligent DSS. The data-driven system applied ML techniques for uncertainty modeling. The data-driven approach can discover various database patterns without relying on prior knowledge while also can handle multiple scenarios and flexible objectives. It can also extract information and knowledge from data without speed [29, 30].

The framework of data-driven optimization under uncertainty could be considered a hybrid system that integrates the data-driven system based on ML to extract useful and relevant information from data. The model-based system is based on mathematical programming to derive the optimal decisions from the information [28]. The inability of traditional optimization methods to analyze big data, as well as recent advances in ML techniques, made data-driven optimization a promising way to hedge against uncertainty in the era of big data and ML. Therefore, these promises create the need for organic integration and effective interaction between ML and mathematical programming. In existing data-driven optimization frameworks, data serve as input to a data-driven system. After that, useful, accurate, and relevant uncertainty information is extracted through the data-driven system and further passed along to the model-based system based on mathematical programming for rigorous and systematic optimization under uncertainty, using paradigms such as robust optimization and stochastic programming.

The various ML techniques and their potentials applications in data-driven optimization under uncertainty are presented in the following.

4.1 Distributionally robust optimization

The stochastic programs are used where the distribution of the uncertain parameters is only observable through a finite training dataset [31]. As the primary assumption in the stochastic programming approach, the probability distribution of uncertain parameters should be clear. However, such complete knowledge of parameters probability distribution is rarely available in practice. In practice, instead of knowing the actual distribution of an uncertainty parameter, what the decision-maker has is a set of historical/ or real-time uncertainty data and possibly some prior structure knowledge of the probability. Also, the assumed possibility distribution of uncertain parameters may deviate from their actual distribution. Moreover, relying on a single probability distribution could lead to sub-optimal solutions or even lead to the deterioration in out-of-sample performance [32]. Motivated by these stochastic programming weaknesses, DRO emerges as a new data-driven optimization paradigm that hedges against the worst-case distribution in an ambiguity set [28]. DRO paradigm integrates data-driven systems and model-based systems. A data-driven approach is applied in the DRO model to construct an uncertainty set of probability distributions from uncertainty data through statistical inference and big data analytics [28]. In data-driven stochastic modeling, the uncertainty is modeled via a family of probability distributions that well capture uncertainty data on hand [28]. This set of probability distributions is referred to as an ambiguity set. With this ambiguity set, a model is then proposed for problem design. Finally, a solution strategy is applied for solving the optimization problem. For example in the literature, the Wasserstein metric has been used, to construct a ball in the space of (multivariate and non-discrete) probability distributions centered at the uniform distribution on the training samples, to seek decisions that perform best in view of the worst-case distribution within this Wasserstein ball [31]. Different practical approaches, such as the moment-based, and the adopted distance metric, were employed for uncertainty constructing [33, 34], and [31]. DRO is an effective method to address the inexactness of probability distributions of uncertain parameters in decision-making under uncertainty that can be applied for optimizing supply chain activities, for planning and scheduling under uncertainty. This way reduces the modeling difficulty for uncertain parameters. Wang & Chen [35] proposed a two-stage DRO model considering scarce data of disasters. A moment-based fuzzy set describes uncertain distributions of blood demand to optimize blood inventory prepositioning and relief activities together. Chiou [36], to regulate the risk associated with hazardous material transportation and minimize total travel cost on the interested area under stochasticity, presented a multi-objective data-driven stochastic optimization model to determine generalized travel cost for hazmat carriers. Gao et al. [37] proposed a two-stage DRO model for better decision making in optimal design and shale gas supply chains under uncertainty. They applied a data-driven approach to construct the ambiguity set based on principal component analysis and first-order deviation functions. In the other study, Ning & You [28] proposed a novel data-driven Wasserstein DRO model for biomass with agricultural waste-to-energy network design under uncertainty. They proposed a data-driven approach to construct the Wasserstein ambiguity set for the feedstock price uncertainty, which is utilized to quantify their distances from the data-based empirical distribution.

4.2 Data-driven robust optimization

A robust optimization is a popular approach for optimization under uncertainty. It defines an uncertainty set of possible realizations of the uncertain parameters and then optimizes against worst-case realizations within this set [5, 6]. In real-world applications, the underlying distribution of uncertainties may be intrinsically complicated and vary under different circumstances [38]. Choosing the accurate underlying distribution of uncertainties and the uncertainty sets by prior knowledge is somewhat challenging in practice. In robust optimization, the uncertainty is formed as an uncertainty set in which any point is a possible scenario [39]. Since the uncertainty set includes the worst case, robust optimization may be over-conservative. It is essential to apply the appropriate approach to construct the uncertainty set and adjust the conservatism level simultaneously [39]. As an essential ingredient in robust optimization, uncertainty sets endogenously determine robust optimal solutions and, therefore, should be devised with special care [28]. However, uncertainty sets in the conventional robust optimization methodology are typically set a priori using a fixed shape and model without providing sufficient flexibility to capture the structure and complexity of uncertainty data [28]. For instance, the geometric shapes of uncertainty set in the conventional robust optimization methodology do not change with the intrinsic structure and complexity of uncertainty data. Furthermore, these uncertainty sets are specified by a finite number of parameters, thereby limiting modeling flexibility. Motivated by this knowledge gap, data-driven robust optimization emerges as a powerful paradigm for addressing uncertainty in decision making.

Choosing a good uncertainty set enables robust optimization models to provide better solutions than other approaches solutions [5, 6]. Poor choice of the uncertainty set makes robust optimization model overly conservative or computationally intractable. In the era of big data, many data are routinely generated and collected containing abundant information about the distribution of uncertainties; thereby, ML tools can construct the uncertainty sets based upon these data. Data-driven robust optimization is a new paradigm for hedging against uncertainty in the era of big data. The ML tools can be applied to estimate data densities with sufficient accuracy and construct an appropriate uncertainty set based upon intelligent analysis and the use of uncertainty data for modeling robust optimization problems. A desirable uncertainty set shall have enough flexibility to adapt to the intrinsic structure behind data, thereby characterizing the underlying distribution and facilitating the solutions.

Data-driven robust optimization could be considered a "hybrid" system that integrates the data-driven system based on ML to construct the uncertainty set from historical uncertainty data. The model-based system is based on the robust programming model to derive the optimal decisions from the information. More specifically, data serves as input to a data-driven system. **Figure 1** presents the data-driven optimization paradigm framework. After that, the data-driven method constructs the uncertainty set to extract information from historical data fully. Constructing the uncertainty sets based upon historical data can be considered as an unsupervised learning problem from an ML perspective. So, data-driven robust optimization is a hybrid system that utilizes ML techniques to design data-driven uncertainty sets and develops a robust optimization problem from the data-driven set. Different effective unsupervised learning models such as the Dirichlet process mixture model, maximum likelihood estimation, principal component analysis, regular and conservative support vector clustering, Bayesian ML, and kernel density estimation were employed for uncertainty constructing, which could provide powerful representations of data distributions [38, 40, 41]. Uncertainty set is the set that can offer robust solutions with a conservatism level. Furthermore, this uncertainty set is finally given to the model-based system based on robust optimization to obtain robust solutions under uncertainty.

ML methods of support vector clustering-based uncertainty set (SVCU) and conservative support vector clustering-based uncertainty set (CSVCU) have been applied to finding an enclosed hypersphere with minimum volume which is able to cover all data samples as tightly as possible as uncertainty sets. Conservative support vector clustering is the most suitable choice for obtaining robust solutions in cases with sufficient data to construct an uncertainty set enclosing future data with a high confidence level [42]. Furthermore, it is the most effective choice for obtaining lower conservative solutions. On the other hand, CSVCU is suitable for highly conservative decision-makers since it is the only set that can offer robust solutions with a high conservatism level, particularly when there is limited data [42]. A data-driven robust optimization under correlated uncertainty was proposed to hedge against the fluctuations generated from continuous production processes in an ethylene plant [43]. For capturing and enrich the valid information of uncertainties, a copula-based method is introduced to estimate the joint probability distribution and simulate mutual scenarios for uncertainties. A deterministic and data-driven robust optimization framework was proposed for energy systems optimization under uncertainty. The uncertainty set is constructed by support vector clustering based on real industrial data [39]. A data-driven robust optimization was applied to design and optimize the entire wastewater sludge to-biodiesel supply chain [42]. They develop a conservative support vector clustering (CSVS) method to construct an uncertainty set from limited data. The developed uncertainty set encloses the fuzzy support neighborhood of data samples, making it practical even when the available data is limited.

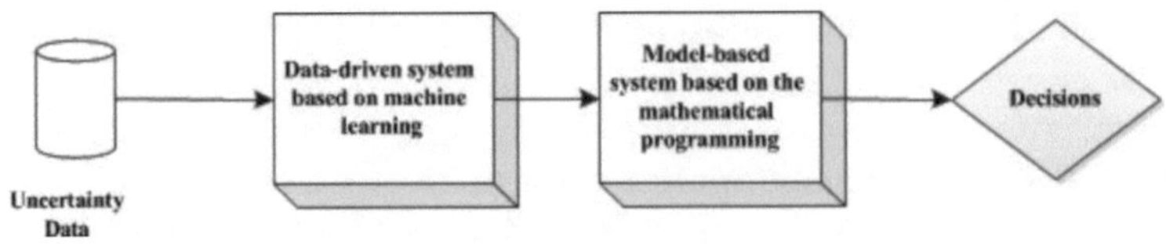

Figure 1.
The schematic of the data-driven optimization paradigm framework.

4.3 Data-driven chance-constrained program

Chance constrained programming is a practical and convenient approach to control risk in decision-making under uncertainty. However, due to unknown probability distributions of uncertainty parameters, the solution obtained from a chance-constrained optimization problem can be biased. In practice, instead of knowing the actual distribution of an uncertainty parameter, only a set of historical/ or real-time uncertainty data, which can be considered as samples taken from the actual (while ambiguous) distribution, can be observed and stored. On the other hand, even if the probability distribution of an uncertainty parameter is available, the chance-constrained program is computationally cumbersome. Motivated by Chance constrained programming weaknesses, data-driven chance-constrained optimization emerges as a new data-driven optimization paradigm. The data-driven stochastic programming approach is a data-driven risk-averse strategy to handle uncertainties in the era of big data effectively.

In contrast to the data-driven stochastic programming approach, data-driven chance-constrained programming is another paradigm focusing on chance constraint satisfaction under the worst-case probability instead of optimizing the worst-case expected objective. Although both data-driven chance-constrained programs and DRO adopt ambiguity sets in the uncertainty models, they have distinct model structures. Specifically, the data-driven chance-constrained program features constraints subject to uncertainty in probability distributions. Simultaneously, DRO typically only involves the worst-case expectation of an objective function concerning a family of probability distributions [28]. In the data-driven stochastic programming approach, historical data is utilized to learn the uncertain parameters' distributions.

Data-driven chance-constrained programs with moment-based ambiguity sets, distance-based ambiguity set, Prohorov metric-based ambiguity sets [44], φ-divergence based ambiguity set [45], kernel smoothing method [46], Wasserstein ambiguity set [47].

Ghosal and Wiesemann [48] applied for Data-driven chance-constrained programs in the capacitated vehicle routing problem (CVRP), which asks for the cost-optimal delivery of a single product geographically dispersed customers through a fleet of capacity-constrained vehicles. They model the customer demands as a random vector whose distribution is only known to belong to an ambiguity set.

4.4 Leveraging DL techniques in the data-driven optimization

The recent development in the data science field, AI, and ML techniques have enabled intelligent and automated DSS and real-time analytics coupled with computing power improvements. Thus, AI techniques are applied to big data sources to extract the knowledge-based rules or identify the underlying rules and patterns by ML techniques, to drive the systems toward set objectives. DL is an ML technique that can extract high levels of information and knowledge from massive data volumes. DL algorithms consist of multiple processing layers to learn representations of data with multiple abstraction levels [26]. For example, recently, DL techniques have been used to accurately forecasting customer demand, price, and inventory leading to optimization of supply chain performance. An intelligent forecasting system leads to optimize performance, reduce costs, and increase sales and profit. DL techniques can apply deep neural network architectures to solve various complex problems. The DL paradigm requires high computing power and a large amount of data for training. The recent advances in parallel architectures and GUP (Graphical Processing Unit) enabled the necessary computing power required in deep neural networks (DNN). The emergence of advanced IoT and blockchain technologies has also solved the need for a large amount of data to learn. IoT and blockchain result in massive amounts of streaming real-time data often referred to as "big data," which brings new opportunities to control and manage supply chains [49]. Optimizing the parameters in DNN is a challenging undertaking. Several optimization algorithms such as Adam, Adagrad, RMSprop, have been proposed to optimize the network parameters in DNN and improve generalizability. This technique, which stabilizes the optimization, paved the way for learning deeper networks [50]. In real applications, uncertainty data exhibit very complex and highly nonlinear characteristics. DNN can be used to uncover useful patterns of uncertainty data for optimizing under uncertainty [28]. Deep data-driven optimization could be considered a "hybrid" system that integrates the deep data-driven system based on DL to forecast the uncertainty parameters. The model-based system is based on mathematical programming to drive the optimal decisions from predicted parameters (the deep data-driven system). In the DL-based system, DNN has been applied to analyze features, complex interactions, and relationships among features of a problem from samples of the dataset and learn model, which can be used for demand, inventory, and price forecasting. Kilimci et al. [51] developed an intelligent demand forecasting system based on the analysis and interpretation of the historical data using different forecasting methods, including support vector regression algorithm, time series analysis techniques, and DL models. In a study, the Auto-Regressive Integrated the backpropagation (BP) network method, recurrent neural network (RNN) method, and Moving Average (ARIMA) model were tested to forecast the price of agricultural products [52]. Yu et al. [53] developed an online big-data-driven forecasting model of Google trends to improve oil consumption prediction. Their proposed forecasting model considers traditional econometric models (LogR and LR) and typical AI techniques (BPNN, SVM, DT, and ELM).

Accurate automatic optimization heuristics are necessary for dealing with the complexity and diversity of modern hardware and software. ML is a proven technique for learning such heuristics, but its success is bound by the quality of the features used. Developers must handcraft these features through a combination of expert domain knowledge and trial and error. This makes the quality of the final model directly dependent on the skill and available time of the system architect. DL techniques are a better way to build heuristics. A deep neural network can learn heuristics over raw code entirely without using code features. The neural network simultaneously constructs appropriate representations of the code and learns how best to optimize, removing the need for manual feature creation. DNN can improve the accuracy of models without the help of human experts. Generally, this approach is a fundamental way to integrate forecast approaches into mathematical optimization models. First, a probabilistic forecast approach for future uncertainties is given by exploiting the advanced DL structures. Second, a model-based system based on mathematical programming is applied to derive the optimal decisions from the forecasting data. Comparison and evaluation of the forecasting models are significant since DL models can have different performances depending on the properties of the data [54, 55]. The performances of DL models differ according to the forecasting time, training duration, target data, and simple or ensemble structure [56, 57].

In a study, Nam et al. [54, 55] applied DL-based models to forecast fluctuating electricity demand and generation in renewable energy systems. This study compares and evaluates DL models and conventional statistical models. The DL models include DNN, long short-term memory, gated recurrent unit, and the disadvantages of conventional statistical models such as multiple linear regression and seasonal autoregressive integrated moving average. In another study, the operation of a cryogenic NGL recovery unit for the extraction of NGL has been optimized by implementing data-driven techniques [58]. The proposed approach is based on an optimization framework that integrates dynamic process simulations with two DL-based surrogate models using a long short-term memory (LSTM) layout with a bidirectional recurrent neural network (RNN) structure. Kilimci et al. [51] developed an intelligent demand forecasting system. This improved model is based on analyzing and interpreting the historical data using different forecasting methods, including time series analysis techniques, support vector regression algorithm, and DL models.

Accessing a sufficient amount of data for some optimization models is a practical challenge. For example, the quality of scenario-based optimization frameworks strongly depends on access to a sufficient amount of uncertain data. However, in practice, the amount of uncertainty data sampled from the underlying distribution is limited. On the other hand, acquiring a sufficient amount of uncertainty data is extremely time-consuming and expensive in some cases, which leads to the limited application of some approaches [59]. To deal with the practical challenge of requiring an insufficient amount of data, deep generative models emerge as a new paradigm to generate synthetic uncertainty data with the aim of better decisions with insufficient uncertainty data. DL techniques could be applied to learn the useful intrinsic patterns from the available uncertainty data and generate synthetic uncertainty data. More specifically, in deep generative models, the correct data distribution is mimicked either implicitly or explicitly by the DL techniques. Then the learned distribution is used to generate new data points referred to as synthetic data [28]. After that, these synthetic data serve as input to an optimizing model to derive the optimal decisions. Some of the most commonly used deep generative models are variational autoencoders generative and adversarial networks [26]. These synthetic uncertainty data generated by the DL techniques can be potentially useful in the scenario-based optimization model.

4.5 Deep data-driven models

DL models are a class of approximate models proven to have strong predictive capabilities for representing complex phenomena [60]. Approximate models are currently experiencing a radical shift due to the advent of DL. However, our research into the existing literature reveals a scarcity of research utilizing DL in approximate modeling. The introduction of DL models into an optimization formulation provides a means to reduce the problem complexity and maintain model accuracy [60]. Recently it has been shown that DL models in the form of neural networks with rectified linear units can be exactly recast as a mixed-integer linear programming formulation. DL is a method to approximate complex systems and tasks by exploiting large amounts of data to develop rigorous mathematical models [60].

Using DNN to model real-world problems is a powerful tool, as they provide an efficient abstraction that can be used to analyze the structure of the task at hand. The rigorous mathematical model is developed based on neural networks modeling complex systems and optimizing their operations in the deep data-driven model framework. This approximate model is developed by exploiting large amounts of data using DL techniques. Then the solving method is applied to obtain the optimal solutions of the developed optimization model. Developing an optimal solution to the approximate model remains challenging [60].

Pfrommer et al. [61] utilized a stochastic genetic algorithm to optimize a composite textile draping process where a neural network was utilized as a surrogate model. Marino et al. [62] presented an approach for modeling and planning under uncertainty using deep Bayesian neural networks (DBNNs). They use DBNNs to learn a stochastic model of the system dynamics. Planning is addressed as an open-loop trajectory optimization problem. In the study, DL-based surrogate modeling and optimization were proposed for microalgal biofuel production and photobioreactor design [63]. This surrogate model is built upon a few simulated results from the physical model to learn the sophisticated hydrodynamic and biochemical kinetic mechanisms; then adopts a hybrid stochastic optimization algorithm to explore untested processes and find optimal solutions. Tang & Zhang [64] developed a deep data-driven framework for modeling combustion systems and optimizing their operations. First, they developed a deep belief network to model the combustion systems. Next, they developed a multi-objective optimization model by integrating the deep belief network-based models, the considered operational constraints, and the control variable constraints.

4.6 Online learning-based data-driven optimization

In conventional data-driven optimization frameworks, a set of uncertainty data serves as input to the data-driven system, in which learning typically takes place once by using learning techniques. This approach fails to account for real-time uncertainty data [28]. For example, in the DRO method, the uncertainty set of probability distributions is constructed from uncertainty data. Once the uncertainty sets of probability distributions are obtained, they remain fixed for the model-based system based on mathematical programming and are not updated or refined. However, in real practice, a vast number of uncertainty data are generated and collected sequentially in an online fashion; therefore, data-driven systems should be developed to analyze the real-time data. An online-learning-based data-driven optimization framework emerges as a new data-driven optimization paradigm. Learning takes place iteratively to account for real-time data, and the data-driven system is updated in an online fashion. The framework of online-learning-based data-driven optimization could be considered a hybrid system that integrates the online data-driven and model-based systems. In the online data-driven system, the real-time uncertainty data should be saved and analyzed sequentially based on ML to extract sequentially useful and relevant information from the real-time data. The online data-driven system (such as the uncertainty sets, probability distributions sets, and forecasting data) that serve as input to a model-based system should be updated in an online fashion. Then in the model-based system, the optimal decisions are made sequentially from the real-time information based on mathematical programming. There is a "feedback" channel for information flow returning from the model-based system to the data-driven system in this framework. The information flow is fed into the mathematical programming problem from the ML results. Using the feedback control strategy delivers amazingly superior system performance (e.g., stability, robustness to disturbances, and safety) [28]. **Figure 2** presents the potential schematic of the online learning-based data-driven optimization system.

The online-learning-based data-driven optimization framework, updating the data-driven systems, and developing efficient algorithms to solve online learning-based mathematical programming problems have become challenging.

4.7 Leveraging RL techniques for optimization

RL has transformed AI, especially after the success of Google DeepMind. This branch of ML epitomizes a step toward building autonomous systems by understanding the visual world. Deep RL is currently applied to different sorts of problems that were previously obstinate. In this subsection, the authors will analyze Deep RL and its applications in optimization.

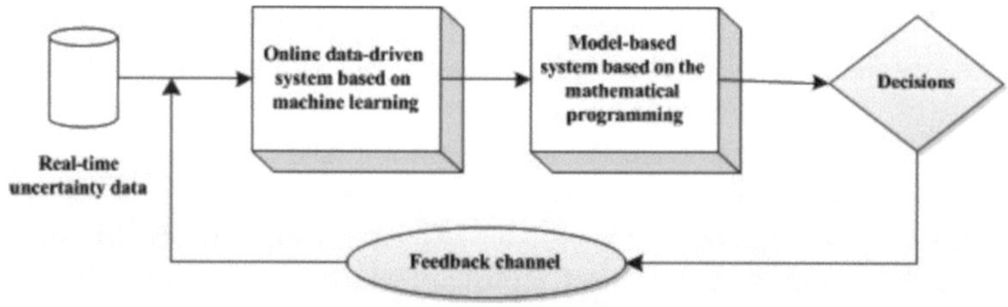

Figure 2.
The schematic of the "closed-loop" online learning-based data-driven optimization framework.

RL is one of the ML areas recently applied to tackle complex sequential decision problems. RL is concerned with how a software agent should choose an action to maximize a cumulative reward. RL is considered an optimal solution in addressing challenges where many factors must be taken into account, like supply chain management. For example, Q-learning is a type of RL algorithm that is applied to tackle simple optimization problems. In this approach, the Q-value has been applied to any state of the system. Although the classical RL algorithms guarantee optimal policy, these algorithms cannot promptly solve large states or actions. Many problems in the real world have large and action spaces. Applying RL algorithms for solving large problems would be nearly impossible, as these models would be costly to train. Therefore, deep RL emerges as a new method in which DNN is used to approximate any of the following RL components. Recently, deep Q-network (DQN) algorithms have been used in different areas. For example, deep Q-network (DQN) algorithms have been applied to solve supply chain optimization problems. These DQNs operate as the decision-maker of each agent. That results in a competitive game in which each DQN agent plays independently to minimize its own cost. Instead, recently a unified framework has been proposed in which the agents still play independently from one another. Still, in the training phase, this model uses a feedback scheme so that the DQN agent learns the total cost for the whole network and, over time, learns to minimize it.

Like other types of reinforcement ML technique, multi-agent RL is a system of agents (e.g., robots, machines, and cars) interacting within a common environment. Each agent decides each time-step and works along with the other agent(s) to achieve a given goal. The agents are learnable units that want to learn policy on the fly to maximize the long-term reward through the interaction with the environment. Recently the multi-agent RL techniques have been applied to develop the supply chain management (SCM) systems that perform optimally for each entity in the chain. A supply chain can be defined as a network of autonomous business entities collectively responsible for procurement, manufacturing, storing, and distribution [65]. Entities in a supply chain have different sets of environmental constraints and objectives.

One of the biggest challenges of the development of MAS based supply chain is designing agent policies. To address designing agent policies, recently, automatic policy designing by RL has drawn attention. RL is considered an optimal solution in addressing challenges where a huge number of factors must be taken into account, like SCM. RL technique does not require datasets covering all environments, constraints, operations, and entity operation results. A multi-agent RL (MARL)-based SCM system can enable agents to learn automatically policies that optimize the supply chain performance using RL concerning certain constraints, environments, and objectives to optimize the performance. More specifically, the RL technique enables an agent to learn a policy by correcting necessary data itself during trial-and-error on the content of operations [66]. All agents also simultaneously cooperate to optimize the performances of the entire supply chain. RL technique can be applied for a certain problem when all processes concerning the problem satisfy a Markov property. Environmental change for a certain agent depends on the previous state of the environment and the agent's action. It is impossible to assume the Markov property because an agent's environmental change depends on the previous state for the agent and the other agent's actions.

There are two problems in developing a MARL technique for SCM: Building Markov decision processes for a supply chain and then avoiding learning stagnation among agents in learning processes. For solving these problems, a learning management method with deep neural network (DNN)-weight evolution (LM-DWE) has been applied [67]. Fuji et al. [67] developed a multi-agent RL technique to develop a supply chain management (SCM) system that enables agents to learn policies that optimize SC performance. They applied a learning management method with deep-neural-network (DNN)-weight evolution (LM-DWE) in the MARL for SCM. An RL framework-FeedRec has been used in a study to optimize long-term user engagement [68]. They used hierarchical LSTM to design the Q-Network to model the complex user behaviors; they also used Q Network to simulate the environment. Zhang et al. [69] proposed a multi-agent learning (MAL) algorithm and applied it for optimizing online resource allocation in cluster networks.

4.8 Deep RL for solving NP-hard problems

Optimization in current DSS has a highly interdisciplinary nature related to integrating different techniques and paradigms for solving complex real-world problems. The design of efficient NP-hard combinatorial optimization problems is a fascinating issue and often requires significant specialized knowledge and trial-and-error. NP-hard problems are solved with exact methods, heuristic algorithms, or a combination of them. Although exact methods provide optimal answers, they have the limitation of performing inefficiently in time complexity. Heuristics are used to improve computational time efficiency and provide decent or near-optimal solutions [70]. According to the definition of Burke et al. [71], a hyper-heuristic is a searching mechanism that aims to select or generate appropriate heuristics to solve an optimization problem. However, the effectiveness of general heuristic algorithms is dependent on the problem being considered, and high levels of performance often require extensive tailoring and domain-specific knowledge. ML strategies have become a promising route to addressing these challenges, which led to the development of meta-algorithms to various combinatorial problems.

Solution approaches meta-heuristics and hyper-heuristics have been developed to tackle the NP-hard combinatorial optimization problem [72]. Recently, hyper-heuristics arise in this context as efficient methodologies for selecting or generating (meta) heuristics to solve NP-hard optimization problems. Hyper-heuristics are categorized into heuristic selection (Methodologies to select) and heuristic generation (Methodologies to generate) [71]. Deep RL is a possible learning method that can automatically solve various optimization problems [73]. Encouragingly, characteristics of the deep RL method have been found in comparison with classical methods, e.g., strong generalization ability and fast solving speed. RL methods can be used at different levels to solve combinatorial optimization problems. They can be applied directly to the problem, as part of a meta-heuristic, or as part of hyper-heuristics [74]. Utilizing advanced computation power with meta-heuristics algorithms and massive-data processing techniques has successfully solved various NP-hard problems. However, meta-heuristic approaches find good solutions which, do not guarantee the determination of the global optimum. Meta-heuristics still face the limitations of exploitation and exploration, which consists of choosing between a greedy search and a wider exploration of the solution space.

A way to guide Meta-heuristic algorithms during the search for better solutions is to generate the initial population of a genetic algorithm by using a technique of Q-Learning algorithm.

The hyper-heuristic for heuristic selection can use RL algorithms, enabling the system to autonomously select the meta-heuristic to use in the optimization process and the respective parameters. For example, Falcão et al. [74] proposed a hyper-heuristic module for solving scheduling problems in manufacturing systems. The proposed hyper-heuristic module uses an RL algorithm, which enables the system to autonomously select the meta-heuristic to use in the optimization process and the respective parameters. Cano-Belmán et al. [75] proposed a heuristic generation scatter search algorithm to address a mixed-model assembly line sequencing problem. Khalil et al. (Dai et al., 2017) developed a neural combinatorial optimization framework that utilizes neural networks and RL to tackle combinatorial optimization problems. The developed meta-algorithm automatically learns good heuristics for a diverse range of optimization problems over graphs. Mosadegh et al. [72] proposed novel hyper-simulated annealing (HSA) to tackle the NP-hard problem. They developed new mathematical models to describe a mixed-model sequencing problem with stochastic processing times (MMSPSP). The HSA applies a Q-learning algorithm to select appropriate heuristics through its search process [72]. The main idea is to conduct simulated annealing (SA)-based algorithms to find a suitable heuristic among available ones creating a neighbor solution(s).

Case study 1: Data-driven robust optimization under correlated uncertainty.

The first case study focuses on the production schedule. The data-driven robust optimization applied for an ethylene plant is predicted to hedge against the fluctuations generated from continuous production processes. For capturing and enrich the valid information of uncertainties, copulas are introduced to estimate the joint probability distribution and simulate mutual scenarios for uncertainties [43]. For this purpose, cutting planes are generated to remove unnecessary uncertain scenarios in the uncertainty sets. Then robust formulations induced by the cut set are proposed to reduce conservatism and improve the robustness of scheduling solutions. They consider the robust counterpart induced by the classical uncertainty set, where the difference to the best possible solution over all scenarios is to be minimized. Instead of focuses on simple uncertainty sets that are either finite or hyperboles, they considered problems with more flexible and realistic ellipsoidal uncertainty sets. In this research, the cut sets of flexible uncertainty sets are proposed. They used the historical data to correct the uncertainties and drive the reformulation of constraints with uncertainties. The new robust formulations induced by cut sets are derived for linear programming (LP) and mixed-integer linear programming (MILP) problems. Through the real-world ethylene plant example, the correlations between uncertain consumption rates of furnaces are analyzed.

In this research, Decision-makers prefer to obtain robust solutions immune to most high-frequency uncertain scenarios. Since in production scheduling problems, many uncertainties are associated with the entire production network, a process, or equipment, which makes them correlated and difficult to be separated. So, in this optimization research, uncertainties are assumed to be dependent. In this research, the cut sets of flexible uncertainty sets are proposed.

Deterministic solutions are regarded as theoretically optimal at most times, and robust solutions provide references for decision-makers, which may not be optimal but feasible and applicable. It is always neglected that stricter descriptions of uncertainties could also create great profits. The full coverage of uncertain values usually leads to unpractical and conservative results. The improper simplification of uncertainty scenarios will cause infeasibility when the solutions are implemented in the volatile production process. Thus, historical data should be introduced to correct the uncertainties and drive the reformulation of constraints with uncertainties. For eliminating the worst-case formulation scenario for robust optimization and decrease conservatism, the cut set of flexible uncertainty sets is constructed by introducing cutting planes. Cutting planes are generated to construct cut sets for the outer approximation of most uncertain scenarios. Since the size of the uncertainty set directly influences the quality of robust solutions, in this research, the more uncertain values are considered.

They stated that utilizing the data-driven robust optimization approach causes the decision-makers to have the ability to decide how many uncertain scenarios are considered in the model and to provide effective, economical, and robust scheduling plans. Finally, it causes fluctuations in the production performance captured and controlled below a lower level of conservatism.

Case study 2: wastewater sludge-to-biodiesel supply chain design.

Designing and optimizing the wastewater sludge-to-biodiesel supply chain facilitates the development of its large-scale production [42]. Hence, this case study evaluates Data-driven robust optimization for supply chain designing and optimization. The entire wastewater sludge-to-biodiesel supply chain over multiple periods is systematically designed and optimized based on the uncertainty sets constructed from the data of uncertain parameters. In this research, a data-driven robust optimization has been adopted, which constructs the uncertainty sets from the data of uncertain parameters utilizing support vector clustering. In contrast, the conventional uncertainty sets are driven without incorporating the data, which results in a high cost of robustness. The developed uncertainty set in this research encloses the fuzzy support neighborhood of data samples that makes it practical even when the available data is limited. The research results show that the proposed data-driven robust optimization approach can yield robust supply chain decisions with the same degree of robustness but at a lower cost than robust conventional optimization approaches.

Case study 3: Forecasting fluctuating variation in electricity demand and generation.

Our third case study relates to forecasting fluctuating electricity demand and generation variation, aiming to develop an energy forecasting model with renewable energy technologies [54, 55]. Wind and solar energy sources are erratic and difficult to implement in renewable energy systems; therefore, circumspection is needed to implement renewable energy systems and policies. This translates into the DL-based models for forecasting fluctuating electricity demand and generation in renewable energy systems.

This study compares and evaluates DL models and conventional statistical models. The DL models include DNN, long short-term memory, gated recurrent unit, and the disadvantages of conventional statistical models such as multiple linear regression and seasonal autoregressive integrated moving average. Thus, they thoroughly compare and evaluate the forecasting models and select the best forecasting model for future electricity demand and renewable energy generation. They then utilized the proposed model for renewable energy scenarios for Jeju Island's policy design to achieve their energy policy. The optimal scenario is assessed by considering its strengths, weaknesses, opportunities, and threats analysis while also considering techno-economic-environmental domestic and global energy circumstances.

5. Conclusion and managerial implications

Data-driven optimization refers to the art and science of integrating the data-driven system based on ML to convert (big) data into relevant and useful information and insights, and the model-based system based on mathematical programming to derive the optimal and more accurate decisions from the information. As a direct implication, the generic approach proposed in data-driven optimization can be utilized to create an automated, data-driven, and intelligent DSS, which would increase the quality of decisions both in terms of efficiency and effectiveness. Recent advances in DL as a predictive model have received great attention lately. One of the distinguishing features of DNN is its ability to "learn" better predictions from large-scale data than ML methods. Hence, one of the primary messages of this overview chapter is to review the applicability of DL in improving DSS across core areas of supply chain operations.

Much data is generated at ever-faster rates by companies and organizations [76]. Applying the advanced DL techniques for predictive analytics becomes a promising issue for further research to improve the decision-making process. Although the conventional data-driven optimization paradigm has made significant progress for hedging against uncertainty, it is foreseeable that data-driven mathematical programming frameworks would proliferate in the next few years due to the generation of large volumes of data and the complexity of relationships among elements. Nowadays, the increase in data acquisition and availability and the emergence of DL makes it imperative to develop data-driven mathematical programming to approximate complex systems under uncertainty. More specifically, a deep data-driven model paradigm, in which the rigorous mathematical model is developed based on neural networks to modeling complex systems and optimizing their operations, could be a promising research direction.

Furthermore, there are some research challenges associated with conventional data-driven optimization frameworks. For example, updating the data-driven system and learning based on real-time data in the data-driven model frameworks can be a key research challenge. Future research could be directed toward designing the data-driven system, in which learning takes place sequentially to extract useful and relevant information from real-time uncertainty data. The data-driven systems should be updated in an online fashion.

Developing the mathematical programming problems for an online-learning-based data-driven optimization paradigm creates another challenge. The model-based system can be devised based on the deep data-driven model paradigm and be leveraged the power of DL. Additionally, deep RL can be applied to developing efficient algorithms to solve the resulting online-learning-based mathematical programming problems. Applying deep RL in the paradigm of learning-while-optimizing also could be another promising research direction. Besides, multi-agent RL techniques could be explored by taking advantage of DL to develop complex systems and optimize their performance based on real-time data.

Also, RL is another ML area that has recently been used to model complex systems and problems and to optimize their performance and behaviors. RL is also considered an optimal solution in addressing challenges where many factors must be taken into account. More specifically, deep RL emerges as a new method to solve the various optimization problems automatically. Thereby, applying RL in optimization problems deserves further attention in future research.

Author details

Saeid Sadeghi[1], Maghsoud Amiri[2] and Farzaneh Mansoori Mooseloo[3]*

1 Faculty of Management and Accounting, Department of Industrial Management, University of Tehran, Tehran, Iran

2 Faculty of Management and Accounting, Department of Industrial Management, Allameh Tabataba'i University, Tehran, Iran

3 Faculty of Management, Department of Industrial Management, University of Hormozgan, Bandar-Abbas, Iran

*Address all correspondence to: farzanmansoori7@gmail.com

References

[1] Biegler, L. T., & Grossmann, I. E. (2004). Retrospective on optimization. *Computers & Chemical Engineering*, *28*(8), 1169-1192.

[2] Sakizlis, V., Perkins, J. D., & Pistikopoulos, E. N. (2004). Recent advances in optimization-based simultaneous process and control design. *Computers & Chemical Engineering*, *28*(10), 2069-2086.

[3] Sahinidis, N. V. (2004). Optimization under uncertainty: state-of-the-art and opportunities. *Computers & Chemical Engineering*, *28*(6-7), 971-983.

[4] Darvazeh, S. S., Vanani, I. R., & Musolu, F. M. (2020). Big data analytics and its applications in supply chain management. In *New Trends in the Use of Artificial Intelligence for the Industry 4.0* (p. 175). IntechOpen.

[5] Bertsimas, D., Gupta, V., & Kallus, N. (2018a). Data-driven robust optimization. *Mathematical Programming*, *167*(2), 235-292.

[6] Bertsimas, D., Gupta, V., & Kallus, N. (2018b). Data-driven robust optimization. *Mathematical Programming*, *167*(2), 235-292.

[7] Grossmann, I. E., Apap, R. M., Calfa, B. A., García-Herreros, P., & Zhang, Q. (2016). Recent advances in mathematical programming techniques for the optimization of process systems under uncertainty. *Computers & Chemical Engineering*, *91*, 3-14.

[8] Birge, J. R., & Louveaux, F. (2011). *Introduction to stochastic programming*. Springer Science & Business Media.

[9] Nikzad, E., Bashiri, M., & Oliveira, F. (2019). Two-stage stochastic programming approach for the medical drug inventory routing problem under uncertainty. *Computers & Industrial Engineering*, *128*, 358-370.

[10] Quddus, M. A., Chowdhury, S., Marufuzzaman, M., Yu, F., & Bian, L. (2018). A two-stage chance-constrained stochastic programming model for a bio-fuel supply chain network. *International Journal of Production Economics*, *195*, 27-44.

[11] Mavromatidis, G., Orehounig, K., & Carmeliet, J. (2018). Design of distributed energy systems under uncertainty: A two-stage stochastic programming approach. *Applied energy*, *222*, 932-950.

[12] Lima, C., Relvas, S., & Barbosa-Póvoa, A. (2018). Stochastic programming approach for the optimal tactical planning of the downstream oil supply chain. *Computers & Chemical Engineering*, *108*, 314-336.

[13] Alipour, M., Zare, K., & Seyedi, H. (2018). A multi-follower bilevel stochastic programming approach for energy management of combined heat and power micro-grids. *Energy*, *149*, 135-146.

[14] Ben-Tal, A., El Ghaoui, L., & Nemirovski, A. (2009). Robust optimization. *Princeton university press*.

[15] Kim, J., Do Chung, B., Kang, Y., & Jeong, B. (2018). Robust optimization model for closed-loop supply chain planning under reverse logistics flow and demand uncertainty. *Journal of cleaner production*, *196*, 1314-1328.

[16] Aalaei, A., & Davoudpour, H. (2017). A robust optimization model for cellular manufacturing system into supply chain management. *International Journal of Production Economics*, 183, 667-679.

[17] Lim, Y. F., & Wang, C. (2017). Inventory management based on target-oriented robust optimization. *Management Science*, *63*(12), 4409-4427.

[18] Vitus, M. P., Zhou, Z., & Tomlin, C. J. (2015). Stochastic control with uncertain parameters via chance constrained control. *IEEE Transactions on Automatic Control*, *61*(10), 2892-2905.

[19] Farina, M., Giulioni, L., & Scattolini, R. (2016). Stochastic linear model predictive control with chance constraints–a review. *Journal of Process Control*, *44*, 53-67.

[20] Guo, Y., Baker, K., Dall'Anese, E., Hu, Z., & Summers, T. H. (2018). Data-based distributionally robust stochastic optimal power flow—Part I: Methodologies. *IEEE Transactions on Power Systems*, *34*(2), 1483-1492.

[21] Carvalho, A., Lefévre, S., Schildbach, G., Kong, J., & Borrelli, F. (2015). Automated driving: The role of forecasts and uncertainty—A control perspective. *European Journal of Control*, *24*, 14-32.

[22] Russell, S., & Norvig, P. (2002). Artificial intelligence: a modern approach.

[23] Ngiam, K. Y., & Khor, W. (2019). Big data and machine learning algorithms for healthcare delivery. *The Lancet Oncology*, *20*(5), e262-e273.

[24] Helm, J. M., Swiergosz, A. M., Haeberle, H. S., Karnuta, J. M., Schaffer, J. L., Krebs, V. E., ... & Ramkumar, P. N. (2020). Machine learning and artificial intelligence: Definitions, applications, and future directions. *Current reviews in musculoskeletal medicine*, *13*(1), 69-76.

[25] Jakhar, D., & Kaur, I. (2020). Artificial intelligence, machine learning and deep learning: definitions and differences. *Clinical and experimental dermatology*, *45*(1), 131-132.

[26] Goodfellow, I., Bengio, Y., Courville, A., & Bengio, Y. (2016). *Deep learning* (Vol. 1, No. 2). Cambridge: MIT press.

[27] Wang, H., Wu, Y., Min, G., Xu, J., & Tang, P. (2019). Data-driven dynamic resource scheduling for network slicing: A deep reinforcement learning approach. *Information Sciences*, *498*, 106-116.

[28] Ning, C., & You, F. (2019). Optimization under uncertainty in the era of big data and deep learning: When machine learning meets mathematical programming. *Computers & Chemical Engineering*, *125*, 434-448.

[29] Wong, A. K. C., & Wang, Y. (2003). Pattern discovery: a data driven approach to decision support. *IEEE Transactions on Systems, Man, and Cybernetics, Part C (Applications and Reviews)*, *33*(1), 114-124.

[30] Yang, H., Jin, Z., Wang, J., Zhao, Y., Wang, H., & Xiao, W. (2019). Data-Driven Stochastic Scheduling for Energy Integrated Systems. *Energies*, *12*(12), 2317.

[31] Esfahani, P. M., & Kuhn, D. (2018). Data-driven distributionally robust optimization using the Wasserstein metric: Performance guarantees and tractable reformulations. *Mathematical Programming*, *171*(1), 115-166.

[32] Smith, J. E., & Winkler, R. L. (2006). The optimizer's curse: Skepticism and postdecision surprise in decision analysis. *Management Science*, *52*(3), 311-322.

[33] Calafiore, G. C., & El Ghaoui, L. (2006). On distributionally robust chance-constrained linear programs. *Journal of Optimization Theory and Applications*, *130*(1), 1-22.

[34] Hu, Z., & Hong, L. J. (2013). Kullback-Leibler divergence constrained distributionally robust optimization. *Available at Optimization Online*. Hyper-heuristics: A survey of the state of the art. *Journal of the Operational Research Society*, *64*(12), 1695-1724.

[35] Wang, C., & Chen, S. (2020). A distributionally robust optimization for blood supply network considering

disasters. *Transportation Research Part E: Logistics and Transportation Review*, *134*, 101840.

[36] Chiou, S. W. (2020). Data-Driven Stochastic Optimization for Transportation Road Network Design Under Uncertainty. In *Handbook of Research on Big Data Clustering and Machine Learning* (pp. 231-278). IGI Global.

[37] Gao, J., Ning, C., & You, F. (2019). Data-driven distributionally robust optimization of shale gas supply chains under uncertainty. *AIChE Journal*, *65*(3), 947-963.

[38] Shang, C., Huang, X., & You, F. (2017). Data-driven robust optimization based on kernel learning. *Computers & Chemical Engineering*, *106*, 464-479.

[39] Shen, W., Li, Z., Huang, B., & Jan, N. M. (2018). Chance-constrained model predictive control for SAGD process using robust optimization approximation. *Industrial & Engineering Chemistry Research*, *58*(26), 11407-11418.

[40] Ning, C., & You, F. (2017). Data-driven adaptive nested robust optimization: general modeling framework and efficient computational algorithm for decision making under uncertainty. *AIChE Journal*, *63*(9), 3790-3817.

[41] Ning, C., & You, F. (2018). Data-driven decision making under uncertainty integrating robust optimization with principal component analysis and kernel smoothing methods. *Computers & Chemical Engineering*, *112*, 190-210.

[42] Mohseni, S., & Pishvaee, M. S. (2020). Data-driven robust optimization for wastewater sludge-to-biodiesel supply chain design. *Computers & Industrial Engineering*, *139*, 105944.

[43] Zhang, Y., Jin, X., Feng, Y., & Rong, G. (2018). Data-driven robust optimization under correlated uncertainty: a case study of production scheduling in ethylene plant. *Computers & Chemical Engineering*, *109*, 48-67.

[44] Erdoğan, E., & Iyengar, G. (2006). Ambiguous chance constrained problems and robust optimization. *Mathematical Programming*, *107*(1), 37-61.

[45] Jiang, R., & Guan, Y. (2016). Data-driven chance constrained stochastic program. *Mathematical Programming*, *158*(1), 291-327.

[46] Calfa, B. A., Grossmann, I. E., Agarwal, A., Bury, S. J., & Wassick, J. M. (2015). Data-driven individual and joint chance-constrained optimization via kernel smoothing. *Computers & Chemical Engineering*, *78*, 51-69.

[47] Ji, R., & Lejeune, M. A. (2021). Data-driven distributionally robust chance-constrained optimization with Wasserstein metric. *Journal of Global Optimization*, *79*(4), 779-811.

[48] Ghosal & Wiesemann, W. (2018). Data-driven chance constrained programs over Wasserstein balls. *arXiv preprint arXiv:1809.00210*.

[49] Khan, P. W., Byun, Y. C., & Park, N. (2020). IoT-Blockchain Enabled Optimized Provenance System for Food Industry 4.0 Using Advanced Deep Learning. *Sensors*, *20*(10), 2990.

[50] Kraus, M., Feuerriegel, S., & Oztekin, A. (2020). Deep learning in business analytics and operations research: Models, applications and managerial implications. *European Journal of Operational Research*, *281*(3), 628-641.

[51] Kilimci, Z. H., Akyuz, A. O., Uysal, M., Akyokus, S., Uysal, M. O., Atak Bulbul, B., & Ekmis, M. A. (2019). An improved demand forecasting model using deep learning approach and proposed decision integration strategy for supply chain. *Complexity*, *2019*.

[52] Weng, Y., Wang, X., Hua, J., Wang, H., Kang, M., & Wang, F. Y. (2019). Forecasting horticultural products price using ARIMA model and neural network based on a large-scale data set collected by web crawler. *IEEE Transactions on Computational Social Systems*, *6*(3), 547-553.

[53] Yu, L., Zhao, Y., Tang, L., & Yang, Z. (2019). Online big data-driven oil consumption forecasting with Google trends. *International Journal of Forecasting*, *35*(1), 213-223.

[54] Nam, K., Hwangbo, S., & Yoo, C. (2020a). A deep learning-based forecasting model for renewable energy scenarios to guide sustainable energy policy: A case study of Korea. *Renewable and Sustainable Energy Reviews*, *122*, 109725.

[55] Nam, K., Hwangbo, S., & Yoo, C. (2020b). A deep learning-based forecasting model for renewable energy scenarios to guide sustainable energy policy: A case study of Korea. *Renewable and Sustainable Energy Reviews*, *122*, 109725.

[56] Li, Q., Loy-Benitez, J., Nam, K., Hwangbo, S., Rashidi, J., & Yoo, C. (2019). Sustainable and reliable design of reverse osmosis desalination with hybrid renewable energy systems through supply chain forecasting using recurrent neural networks. *Energy*, *178*, 277-292.

[57] Loy-Benitez, J., Vilela, P., Li, Q., & Yoo, C. (2019). Sequential prediction of quantitative health risk assessment for the fine particulate matter in an underground facility using deep recurrent neural networks. *Ecotoxicology and environmental safety*, *169*, 316-324.

[58] Zhu, W., Chebeir, J., & Romagnoli, J. A. (2020). Operation optimization of a cryogenic NGL recovery unit using deep learning based surrogate modeling. *Computers & Chemical Engineering*, *137*, 106815.

[59] Gupta, V., & Rusmevichientong, P. (2017). Small-data, large-scale linear optimization with uncertain objectives. *Management Science*, *67*(1), 220-241.

[60] Katz, J., Pappas, I., Avraamidou, S., & Pistikopoulos, E. N. (2020). Integrating deep learning models and multiparametric programming. *Computers & Chemical Engineering*, *136*, 106801.

[61] Pfrommer, J., Zimmerling, C., Liu, J., Kärger, L., Henning, F., & Beyerer, J. (2018). Optimisation of manufacturing process parameters using deep neural networks as surrogate models. *Procedia CiRP*, *72*, 426-431.

[62] Marino, D. L., & Manic, M. (2019). Modeling and planning under uncertainty using deep neural networks. *IEEE Transactions on Industrial Informatics*, *15*(8), 4442-4454.

[63] del Rio-Chanona, E. A., Wagner, J. L., Ali, H., Fiorelli, F., Zhang, D., & Hellgardt, K. (2019). Deep learning-based surrogate modeling and optimization for microalgal biofuel production and photobioreactor design. *AIChE Journal*, *65*(3), 915-923.

[64] Tang, Z., & Zhang, Z. (2019). The multi-objective optimization of combustion system operations based on deep data-driven models. *Energy*, *182*, 37-47.

[65] Swaminathan, J. M., Smith, S. F., & Sadeh, N. M. (1998). Modeling supply chain dynamics: A multiagent approach. *Decision sciences*, *29*(3), 607-632.

[66] Mnih, V., Kavukcuoglu, K., Silver, D., Rusu, A. A., Veness, J., Bellemare, M. G., ... & Hassabis, D. (2015). Human-level control through deep reinforcement learning. *nature*, *518*(7540), 529-533.

[67] Fuji, T., Ito, K., Matsumoto, K., & Yano, K. (2018, January). Deep multi-agent reinforcement learning using

dnn-weight evolution to optimize supply chain performance. In *Proceedings of the 51st Hawaii International Conference on System Sciences*.

[68] Zou, L., Xia, L., Ding, Z., Song, J., Liu, W., & Yin, D. (2019, July). Reinforcement learning to optimize long-term user engagement in recommender systems. In *Proceedings of the 25th ACM SIGKDD International Conference on Knowledge Discovery & Data Mining* (pp. 2810-2818).

[69] Zhang, C., Lesser, V. R., & Shenoy, P. J. (2009, July). A Multi-Agent Learning Approach to Online Distributed Resource Allocation. In *Ijcai* (Vol. 9, pp. 361-366).

[70] Dumitrescu, I., & Stützle, T. (2003, April). Combinations of local search and exact algorithms. In *Workshops on Applications of Evolutionary Computation* (pp. 211-223). Springer, Berlin, Heidelberg.

[71] Burke, E. K., Gendreau, M., Hyde, M., Kendall, G., Ochoa, G., Özcan, E., & Qu, R. (2013).

[72] Mosadegh, H., Ghomi, S. F., & Süer, G. A. (2020). Stochastic mixed-model assembly line sequencing problem: Mathematical modeling and Q-learning based simulated annealing hyper-heuristics. *European Journal of Operational Research*, *282*(2), 530-544.

[73] Li, K., Zhang, T., & Wang, R. (2020). Deep reinforcement learning for multi-objective optimization. *IEEE transactions on cybernetics*.

[74] Falcão, D., Madureira, A., & Pereira, I. (2015, June). Q-learning based hyper-heuristic for scheduling system self-parameterization. In *2015 10th Iberian Conference on Information Systems and Technologies (CISTI)* (pp. 1-7). IEEE.

[75] Cano-Belmán, J., Ríos-Mercado, R. Z., & Bautista, J. (2010). A scatter search based hyper-heuristic for sequencing a mixed-model assembly line. *Journal of Heuristics*, *16*(6), 749-770.

[76] Corbett, C. J. (2018). How sustainable is big data?. *Production and Operations Management*, *27*(9), 1685-1695.

Chapter 9

Practical Application Using the Clustering Algorithm

Yoosoo Oh and Seonghee Min

Abstract

This chapter will survey the clustering algorithm that is unsupervised learning among data mining and machine learning techniques. The most popular clustering algorithm is the K-means clustering algorithm; It can represent a cluster of data. The K-means clustering algorithm is an essential factor in finding an appropriate K value for distributing the training dataset. It is common to find this value experimentally. Also, it can use the elbow method, which is a heuristic approach used in determining the number of clusters. One of the present clusterings applied studies is the particulate matter concentration clustering algorithm for particulate matter distribution estimation. This algorithm divides the area of the center that the fine dust distribution using K-means clustering. It then finds the coordinates of the optimal point according to the distribution of the particulate matter values. The training dataset is the latitude, longitude of the observatory, and PM10 value obtained from the AirKorea website provided by the Korea Environment Corporation. This study performed the K-means clustering algorithm to cluster feature datasets. Furthermore, it showed an experiment on the K values to represent the cluster better. It performed clustering by changing K values from 10 to 23. Then it generated 16 labels divided into 16 cities in Korea and compared them to the clustering result. Visualizing them on the actual map confirmed whether the clusters of each city were evenly bound. Moreover, it figures out the cluster center to find the observatory location representing particulate matter distribution.

Keywords: clustering, machine learning, data mining, K-means clustering, particulate matter

1. Introduction

This chapter introduces the data mining and the clustering algorithm, which is unsupervised learning among machine learning techniques. In this chapter, we analyze the performed clustering application research that used the air pollution concentration data. It has been a problem recently. The most popular algorithm among the clustering is the K-means clustering algorithm; it represents a data cluster. It is an essential factor that finds an appropriate K value for the distribution of the training dataset. Commonly, we determine the K value experimentally, and at this point, we can set the value using the elbow technique.

One example of clustering application studies is the air pollution concentration clustering algorithm. Air pollution is a substance that causes respiratory diseases and cancer, and the WHO reported the severity of the particulate matter [1–3]. The Korean government has also started providing particulate matter and air pollution information since 2004. On the AirKorea website, we can obtain air pollution information measured at 353 observatories in real-time [4].

Currently, observatories of air pollution in Korea are mainly located in Seoul and Gyeonggi-do, so it is challenging to know accurate air pollution values in local small towns without observatories. Therefore, in this chapter, we study the clustering method for air pollution observatory according to the air pollution concentration. We first split the air pollution-centered regions that can predict the distribution of air pollution by using K-means clustering. Then, we find the optimal station location according to the distribution of air pollution concentrations. Based on the optimal location, we divide the territory of the Korean.

We collect air pollution data in April 2020 and label air pollution monitoring stations through clustering algorithms for this clustering study. Based on the cluster center point, we can apply the Voronoi algorithm to divide the territory of Korea. With this method, we can classify air pollution areas by considering the concentration distribution of air pollution, unlike traditional administrative districts. Furthermore, this method can help know the air pollution distribution in the shaded area without air pollution [5, 6].

2. Related works

In this section, we analyzed related studies to predict the concentration of fine dust [7–12]. The related studies use air pollution data and meteorological data together. In particular, the accuracy of prediction is high when weather data such as temperature and wind speed are used rather than air pollution data [7]. Traditionally, the studies predict the concentration of fine dust through machine learning methods such as linear regression or support vector regression. However, these methods are challenging to consider the spatiotemporal correlation [8]. Therefore, it focuses on improving prediction accuracy by using deep learning [9–12]. There are four distinct seasons in Korea depending on the air mass, so there is a significant difference in the concentration of fine dust by season. Therefore, we must be considered the relationship between location and time.

Joun et al. predicted the concentration of fine dust using the MLR, SVR, ARIMA, and ARIMAX [11]. In this paper, the training datasets are air pollution data (NO2, SO2, CO, O3, PM10) and meteorological data (temperature, precipitation, wind speed). They confirmed that time, location, NO2, CO, O3, SO2, maximum temperature, precipitation, and maximum wind speed were significant variables using multiple linear regression analysis. In addition, they used multiple linear regression and support vector regression to predict fine dust distribution. The prediction accuracy was higher in the artificial neural network than in the multiple support vector regression. If the PM10 concentration increased above 100, the support vector regression was exceptionally high. They perform experiments using ARIMA and ARIMAX to analyze the factors of time according to the location. As a result, there was a difference in the learning accuracy according to the location of the experimental data. Furthermore, the accuracy was higher in using the air quality factor and the meteorological factor than using only the time variable.

Cho et al. designed a predictive model through multiple linear regressions and artificial neural networks and performed the fine-dust prediction [12]. They collected the training data, air pollution data (NO2, SO2, CO, O3, PM10), and meteorological data (temperature, humidity, wind speed, wind direction).

As a result of analyzing the errors by performing prediction, the accuracy of the prediction model using artificial neural networks was better overall than that of multiple linear regression. As the result of the experiment by changing the hidden layers of the artificial neural network, the performance of the multi-layer perceptron was better when there were three hidden layers.

3. Air pollution data mining

The algorithm for air pollution concentration clustering performs clustering through observatories' location and measurement values. We can utilize air pollution information in AirKorea provided by the Korea Environment Corporation. The values of NO2, SO2, CO, O3, PM10, PM25 are upload every hour [13].

In this chapter, we use air pollution information measured in April 2020. First, we download national data on a website to create a feature dataset. And then, we should be converted the address of the observatory to latitude and longitude coordinates of the WGS84 coordinate system. We used Kakao Map API [14]. **Figure 1** shows the used dataset. For example, the 1 row is air pollution code is 111121, a date is April 1, 2020, and the location is the nearby city hall of Seoul.

The feature dataset used for air pollution clustering is latitude and longitude, NO2, SO2, CO, O3, PM10, PM25. We calculate an average observatory data to make one-day data into 1-hour data. Also, we filled in the missing values for each station by obtaining the value of the other stations closest to it. Pseudocode 1 is performing this process.

```
SET myData to READ(fileName)
SET feature to ['latitude', 'longitude', 'PM10', 'SO2', 'CO', 'O3', 'NO2', 'PM25']
IF type of 'date' in myData is string THEN
   convert datetime type to string type of 'date'
ELSE
   PASS
ENDIF
SET group to 'feature data' by 'date' in 'station code'
CALCULATE AVERAGE 'feature data' by group
```

Pseudocode 1.
The process of making the dataset by the day.

4. Air pollution area division method

In this section, we perform location labeling through K-means clustering using created air pollution dataset. The K-means clustering algorithm can be partitioning all data by defining the number of clusters and obtaining the center point of clusters. We used the Scikit-learn python library [15].

Station Code	Date	SO2	CO	O3	NO2	PM10	PM25	Latitude	Longitude
111121	2020-04-01 00:00:00	0.003435	0.443478	0.039565	0.015826	61.0869565	34	37.56455491	126.975615
111122	2020-04-01 00:00:00	0.003609	0.508696	0.035304	0.023826	71.3478261	32.9565217	37.55483769	126.9717341
111123	2020-04-01 00:00:00	0.004619	0.5	0.037095	0.017571	61.8571429	29.7619048	37.57206081	127.0050305
111124	2020-04-01 00:00:00	0.004261	0.578261	0.044304	0.023864	70.0869565	37.2608696	37.56863712	126.9981844
111125	2020-04-01 00:00:00	0.003783	0.473913	0.044783	0.01813	73	32.6956522	37.5709184	126.9965543
111131	2020-04-01 00:00:00	0.003524	0.504762	0.040952	0.020524	71	29.8095238	37.54043943	127.0042997
111141	2020-04-01 00:00:00	0.003217	0.513043	0.040652	0.018913	78.9130435	33.5454545	37.54655201	127.0921031
111142	2020-04-01 00:00:00	0.003783	0.46087	0.044565	0.02187	63.6521739	30.3043478	37.54307024	127.0417993
111143	2020-04-01 00:00:00	0.004087	0.430435	0.030522	0.039522	78.5217391	27.2173913	37.53895452	127.0416703
111151	2020-04-01 00:00:00	0.003632	0.510526	0.043842	0.021842	70	31.2631579	37.58491128	127.0940388
111152	2020-04-01 00:00:00	0.003913	0.513043	0.044391	0.020826	61.9130435	28.0434783	37.57641471	127.0283873
111154	2020-04-01 00:00:00	0.003957	0.552174	0.042217	0.031304	66.4347826	31.7391304	37.58045345	127.0443697

Figure 1.
The air pollution dataset in April 2020.

Because it has to be performed clustering by calculating the distance between each data, we normalize to a value between 0 and 1 using MinMaxScaling [15]. Moreover, we perform the K-means clustering using normalized data. Eq. (1) shows the dataset that maximizes the degree of cohesion within each set. The distance between it in each cluster is measured using the Euclidean distance. The clustering algorithm ends when the average value of this distance no longer changes.

$$\arg\min \sum_{i=1}^{k} \sum_{x \in S_i} \|x - \mu_i\|^2 \tag{1}$$

We calculate the inertia value to find the appropriate K value for K-means Clustering. The inertia value is the sum of the distances between clusters at each center point after clustering. **Figure 2** shows the inertia value according to the K. The optimal k value is where the inertia value decreases rapidly, and the change is not significant. However, it is difficult to determine the optimal k value in this graph. Therefore, we set the k value to 16, focusing on dividing the whole country into 16 provinces.

```
SET myData to READ(fileName)
SET feature to ['latitude', 'longitude', 'PM10', 'SO2', 'CO', 'O3', 'NO2', 'PM25']
FOR day to 31 DO
   GET 'feature' data in myData on 'day'
   CALL MinMaxScaler()
   CALL k-MeansClustering(k is 16)
```

Pseudocode 2.
The process of performing scaling and clustering.

Pseudocode 2 is the source code that loads April data, performs the scaling and clustering. Also, **Figure 3** presents the coordinates of the center point of each cluster as a result of performing clustering based on the air pollution data for a month.

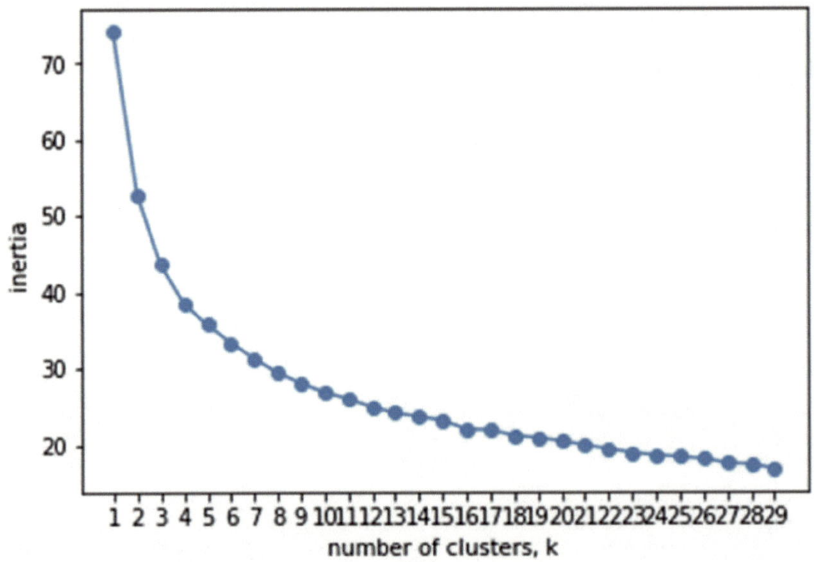

Figure 2.
The inertia value according to the K.

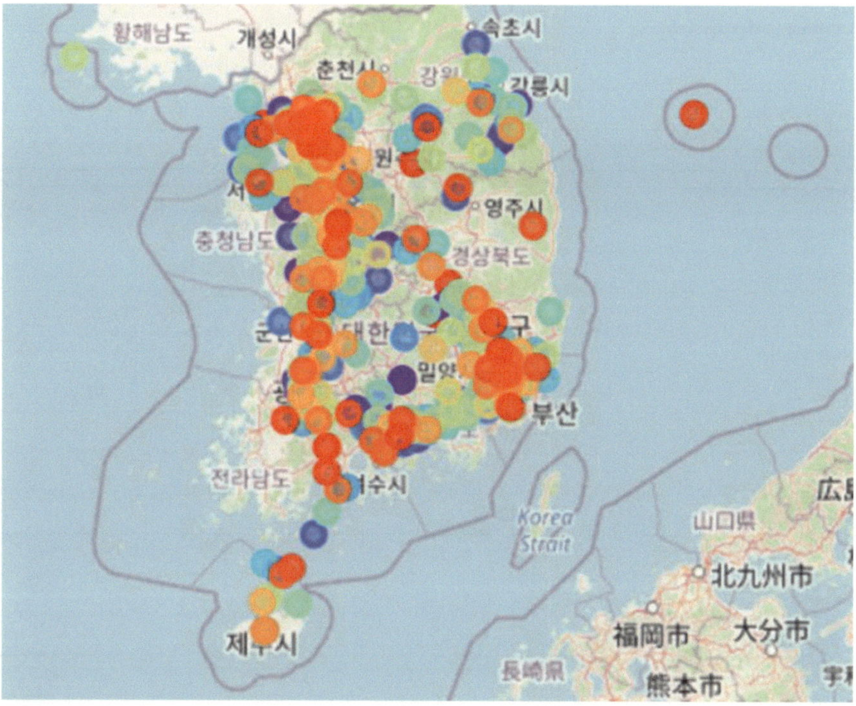

Figure 3.
The coordinates of the center point of each cluster for a month.

We use the Folium python library to show this map [16]. The marker of the same color is the cluster's point divided into 16 in the cluster for the day. Also, **Table 1** is a comparison of 16 administrative district labels and clustering results. For example, the 0 label is the Gangwon-do area, and 11, 12, 15 labels contain the twelve air pollution stations in this district.

To determine the coordinates of the 16 cluster's center point, we perform the K-means clustering again. **Figure 4** is the visualization of the 16 center points on a map to divide regions. As a result of performing clustering, it is more closely distributed in the Seoul cluster center, Incheon, and Gyeonggi-do than other regions. It is because many air pollution monitoring stations are mainly distributed in the metropolitan area in Korea.

Figure 5 shows the results of classifying air pollution monitoring stations by calculating the distance to each station from the obtained 16 center coordinates. Points on the map are the location of the air pollution monitoring station. In this case, we calculated the Euclidean distance using latitude and longitude.

Also, **Figure 6** visualizes the convex hull polygon by connecting the outermost point of the classified measurement stations as a line [17]. This method has the advantage of accurately classifying even if the distance between each point is close because classification is performed based on the location of the stations. However, in an area without an observatory, it is a shaded area, and the distribution of air pollution cannot be measured.

This chapter found cluster's center points using the location and concentration of air pollution monitoring stations to divide air pollution areas that can reflect data distribution. The stations are classified based on the center coordinates, and air pollution areas are divided using the Convexhull polygon. However, there was a problem that the classified air pollution areas did not include areas without air pollution monitoring stations.

Administrative District	Clustering Label															
	0	**1**	**2**	**3**	**4**	**5**	**6**	**7**	**8**	**9**	**10**	**11**	**12**	**13**	**14**	**15**
Gangwon-do												3	4			5
Gyeonggi-do	22			11			38						2		7	
Gyeongsangnam-do		1	6						1					9		
Gyeongsangbuk-do		5						5	1		1	5		3		
Gwangju			1		8											
Daegu		9						3			1			1		
Deajeon						6			3						1	
Busan								7			4			10		
Seoul	8			10			21									
Ulsan								5			9			2		
Incheon	3			12												
Jeollanam-do			8		8					3						
Jeollabuk-do					3	2			14							
Jeju-do										5						
Chungcheongnam-do	1					15			1						6	
Chungcheongbuk-do						6							6		1	1

Table 1.
Number of stations in each cluster by administrative district.

Figure 4.
The visualization of the 16 center points on a map to divide regions.

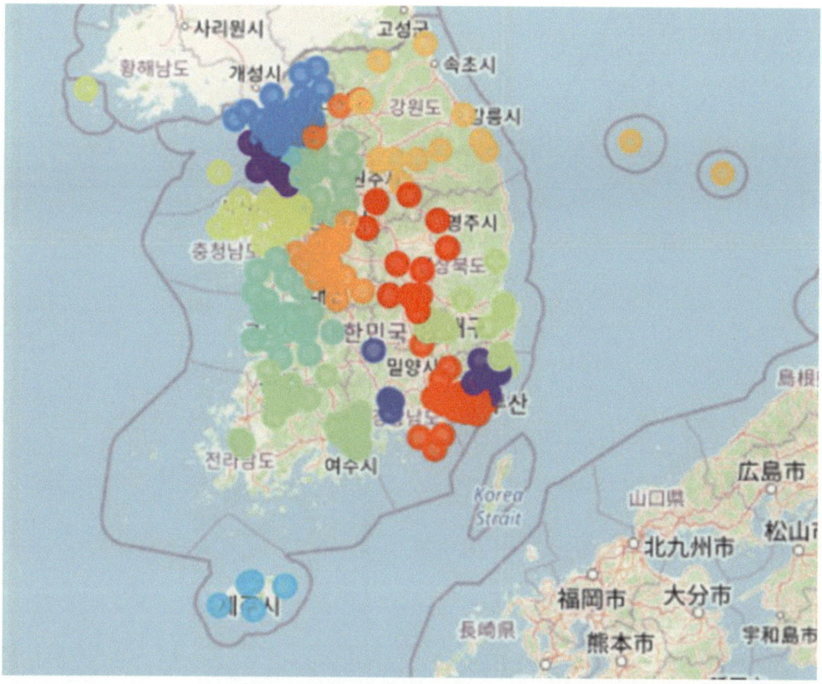

Figure 5.
The results of classifying air pollution monitoring stations by cluster.

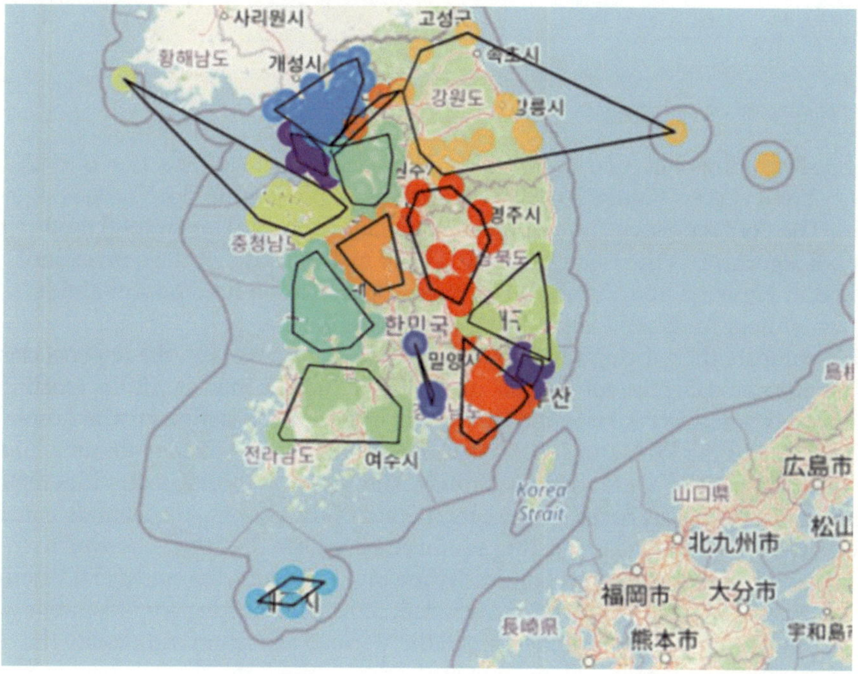

Figure 6.
The result using the convex hull polygon algorithm.

Therefore, we use the Voronoi algorithm to include areas without measurement stations [18]. Also, it can classify areas based on the center point of the cluster. The Voronoi algorithm is to get a line segment that can divide the distance between

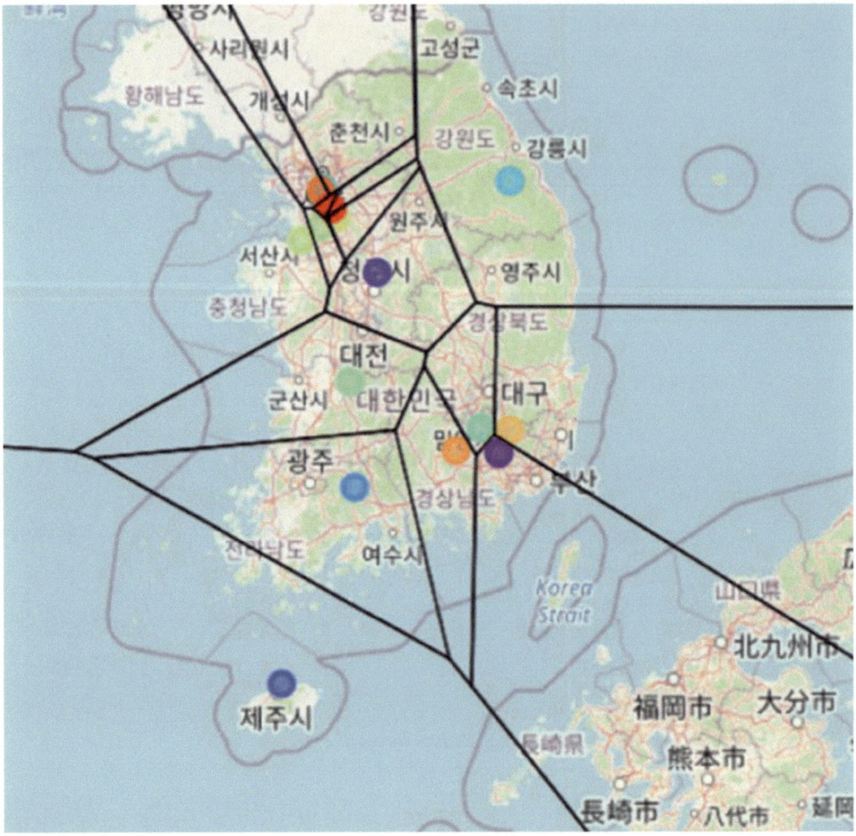

Figure 7.
The result using the Voronoi algorithm.

neighboring points into two and obtain a polygon with the intersection of each line segment as a vertex. **Figure 7** shows the divided regions using the Voronoi algorithm. The dots represent the centers of classified clusters. The method used in the Voronoi algorithm is the Euclidean calculation method. Unlike the convex hull method in **Figure 6**, the Voronoi algorithm's classification method can divide regions without shadowed areas of the Korean Peninsula.

We compare the existing administrative districts in Korea [19], the regional classification method using the convex hull method, and the Voronoi algorithm. Existing administrative districts are classified according to the criteria defined in the Administrative District Practice Manual. Also, the convex hull method divided the area into classified air pollution measurement stations. The Voronoi algorithm classifies regions using the distance value based on the center point of the cluster. Air pollution concentrations were not reflected in existing administrative districts, but the convex hull method and Voronoi algorithm can classify regions. Finally, in the convex hull method, the area without a measuring station is shaded, unlike the existing administrative area and Voronoi algorithm. Comprehensively, the Voronoi algorithm can classify the region by reflecting the air pollution concentration without the shaded area.

5. Conclusion

In this chapter, we collected the data of air pollution stations in Korea and used K-means clustering to learn about data mining and machine learning algorithms.

We divide air pollution areas to predict the distribution of air pollution using air pollution concentration clustering. The training dataset is latitude, longitude, NO2, SO2, CO, O3, PM10, PM25, with air pollution data for one month in April 2020. We use the collected dataset and classify air pollution monitoring stations. Based on the central coordinates of the cluster, the areas of the Korean territory were classified through the Voronoi algorithm. Finally, we confirmed that the proposed air pollution area could be classified by considering the distribution of air pollution, unlike traditional administrative districts. Moreover, the proposed area can help understand the distribution of air pollution in the shaded areas that do not have air pollution stations.

Acknowledgements

This research was supported by the Daegu University, 2018.

Author details

Yoosoo Oh* and Seonghee Min
Daegu University, Gyeongsan-si, Republic of Korea

*Address all correspondence to: yoosoo.oh@daegu.ac.kr

References

[1] WHO, Air pollution, May 2018, Available from: https://www.who.int/news-room/fact-sheets/detail/ambient-(outdoor)-air-quality-and-health/ [Accessed: 2021-06-01]

[2] Hänninen O. O. WHO Guidelines for Indoor Air Quality: Dampness and Mold. In Fundamentals of mold growth in indoor environments and strategies for healthy living. Wageningen Academic Publishers, Wageningen. 2011. p. 277-302.

[3] World Health Organization, WHO air quality guidelines global update, report on a working group meeting, Bonn, Germany, 18–20 October, 2005.

[4] Air Korea, Available from: http://www.airkorea.or.kr/ [Accessed: 2021-06-01]

[5] Min S, Oh, Y. A Study of Particulate Matter Clustering for PM10 Distribution Prediction, In: Proceedings of the International Symposium on Innovation in Information Technology and Applications (2019 ISIITA); 11-13 February 2019; Okinawa. p. 53-56.

[6] Min S, Oh Y. A study of particulate matter area division using PM10 data clustering: Focusing on the case of Korean particulate matter observatory. Journal of Adv Research in Dynamical and Control Systems. 2019;11.12: 959-965. DOI:10.5373/JARDCS/V11SP12/20193300

[7] Munir S, Habeebullah TM, Seroji AR, Morsy EA, Mohammed AM, Saud, WA, Awad AH. Modeling particulate matter concentrations in Makkah, applying a statistical modeling approach. Aerosol Air Quality Research. 2013;13.3:901-910.

[8] Li X, Peng L, Hu Y, Shao J, Chi T. Deep learning architecture for air quality predictions. Environmental Science and Pollution Research. 2016; 23.22:22408-22417.

[9] Freeman BS, Taylor G, Gharabaghi B, Thé J. Forecasting air quality time series using deep learning. Journal of the Air and Waste Management Association. 2018;68.8: 866-886.

[10] Qi Z, Wang T, Song G, Hu W, Li X, Zhang Z. Deep air learning: Interpolation, prediction, and feature analysis of fine-grained air quality. IEEE Transactions on Knowledge and Data Engineering. 2018;30.12: 2285-2297.

[11] Joun S, Choi J, Bae J. Performance Comparison of Algorithms for the Prediction of Fine Dust Concentration. In: Proceedings of Korea Software Congress 2017, 8-10 February 2019; Pyeong Chang. p. 775-777.

[12] Cho K, Jung Y, Kang C, Oh C. Conformity assessment of machine learning algorithm for particulate matter prediction. Journal of the Korea Institute of Information and Communication Engineering. 2019;23.1: 20-26.

[13] AirKorea, Available from: http://www.airkorea.or.kr/ [Accessed: 2021-06-01]

[14] Kakao Map API, Available from: https://apis.map.kakao.com/ [Accessed: 2021-06-01]

[15] Scikit-learn, Available from: https://scikit-learn.org/ [Accessed: 2021-06-01]

[16] Folium Python, Available from: https://python-visualization.github.io/folium/ [Accessed: 2021-06-01]

[17] Kirkpatrick DG, Seidel, R. The ultimate planar convex hull algorithm?. SIAM journal on computing. 1986;15.1: 287-299.

[18] Fortune S. A sweepline algorithm for Voronoi diagrams. Algorithmica. 1987;2.1:153-174.

[19] 2011 Administrative Manual, Ministry of the Interior and Safety, Available from: https://www.mois.go.kr/frt/bbs/type001/commonSelectBoardArticle.do?bbsId=BBSMSTR_000000000055&nttId=77460 [Accessed: 2021-06-01]

Chapter 10

Leaching Mechanisms of Trace Elements from Coal and Host Rock Using Method of Data Mining

Yao Shan

Abstract

Coal and host rock, including the gangue dump, are important sources of toxic elements, which have high-contaminating potential to surface and groundwater. Surface water in the coal mine area and groundwater in the active or abandoned coal mines have been observed to be polluted by trace elements, such as arsenic, mercury, lead, selenium, cadmium. It is helpful to control pollution caused by the trace elements by understanding the leaching behavior and mechanism. The leaching and migration of the trace elements are controlled mainly by two factors, trace elements' occurrence and the surrounding environment. The traditional method to investigate elements' occurrence and leaching mechanism is based on the geochemical method. In this research, the data mining method was applied to find the relationship and patterns, which is concealed in the data matrix. From the geochemical point of view, the patterns mean the occurrence and leaching mechanism of trace elements from coal and host rock. An unsupervised machine learning method, principal component analysis was applied to reduce dimensions of data matrix of solid and liquid samples, and then, the re-calculated data were clustered to find its co-existing pattern using the method of Gaussian mixture model.

Keywords: coal, host rock, occurrence, principal component analysis, Gaussian mixture model

1. Introduction

Coal is a complex system, which contains most elements in the periodic table. The origin of the coal was organic matter containing virtually every element in the periodic table, mainly carbon, but also trace elements. The elements with relative higher content in the coal and host rock, such as iron (Fe) and aluminum (Al), which usually take 1–20% of the rock, respectively, and sodium (Na), potassium (K), calcium (Ca), magnesium (Mg), which are usually in the range of 0.01–10% of the rock, respectively. The trace elements refer to the elements at the 10–10,000 ppm levels in coal, rocks, and soil, etc. A variety of chemicals are associated with coal that is either found in the coal or in the rock layers that lie above and beneath the seams of coal [1]. Some of the trace elements are of great health concern. For example, lead (Pb) accounts for most of the cases of pediatric heavy metal poisoning and makes it difficult for children to learn, pay attention, and succeed in school. Mercury (Hg) exposure puts newborns at risk of neurological deficits and increased cardiovascular risk in adults. Arsenic (As) could cause heavy metal poisoning in adults and does not leave the body once it enters.

Coal mining has caused global environmental concern due to mainly two reasons—first, the coal and host rock contains multiple kinds of toxic trace elements, some of which are of great environmental and health issues, most of them (As, Cd, Co, Cr, Cu, Mn, Ni, Pb, Se, Sn V, and Zn) are associated with inorganic matter [2, 3]; second, the trace elements may be released through combustion and water-rock interaction [3–9].

The coal mine water, containing toxic trace elements, has influenced the water quality of both the groundwater and surface water in China. To control the contamination of trace elements, a lot of efforts have been making in both research and management. According to the Chinese national standard GB/T 19223-2015, the coal mine water is defined as bursting water, infiltrating water from surface water, and working produced water, during coal mining activity. The water is classified into acid ($pH < 6.0$), neutral (6.0–9.0) and alkaline ($pH > 9$), low- (<1000 mg/L), medium- (1000–6000 mg/L), and high-mineralized water (>6000 mg/L), and low- (<50 mg/L), medium- (50–500 mg/L), and high-suspended (>500 mg/L) coal mine water, regarding pH value, total dissolved solids, and suspended matter, respectively. Trace elements released from the coal and rock may contaminate surface and groundwater, including selenium (Se), As, Pb, fluorine (F), Hg, etc., leading to some different unique characteristics of the coal mine water. However, the releasing patterns are relatively similar among the coal mine waters. In the coal-bearing seam, the primitive environment is H-rich and reductive, where some reductive minerals are stable, such as pyrite, chalcopyrite, and sphalerite. While the coal and rock seam contact with air, the Eh value of the surrounding environment is elevated, and the minerals are oxidized [10, 11]. Through this process, the pH value may be reduced, accompanying the release of metal elements into the water, and high concentrations of metal trace elements in the water [12–14]. However, the neutral and alkaline mine water is also common, because of the dissolution of alkaline minerals, such as calcite and dolomite. The net effect of which determines the pH value of coal mine water produces a high mineralization value [12, 15].

Besides the water parameters, the occurrence of trace elements also influences its migration [16–19]. Main minerals in coal include quartz, clay, sulfur-contained minerals, and a lesser number of feldspars and carbonates [20, 21]. As, Cr, Pb, Hg, Mo, Zn, and Sb were found to be enriched in coal compared with continent crust [22–25], while compared to coal, host rock and gangue rejected on the land of coal can release up to 10 times toxic elements into water [2, 26–28].

The migration behavior of trace elements is controlled by two factors, the trace element occurrence and the surrounding environment. However, migration patterns and mechanism of trace elements into a surrounding water body are complex and different depending on the investigating sites. Traditional methods to investigate this process are based on geochemical surveys and testing. The information and pattern behind the data matrix are hard to identify. Along with the development of machine learning, multivariate analytical technology has been applied in some different areas of the geochemical research, the fourth paradigm for the research is becoming a more and more powerful tool to find a solution among the mass data. The multivariate analysis has been used to study the water characteristics [29], source [30, 31], groundwater pathway [32–34], etc. By using the method of multivariate technology, it is possible to disclose the leaching mechanism from the view of trace element occurrence and leaching behavior.

2. Applications of multivariate analysis in geochemistry

The geochemical issues involve a sample-parameter matrix, which includes a co-existence pattern among the parameters and samples. It is cumbersome and hard to identify the patterns using traditional geochemical technology. Thanks to the technological development of artificial intelligence, and the technique of machine learning, the multivariate parameter problem could be solved or mined to discover knowledge or criteria. In the field of geochemistry, the problems are feasible to be solved by using the multivariate analysis method. The multivariate analysis method can be classified to be supervised, unsupervised, and semi-supervised, depending on whether the target parameters are labeled. The unsupervised algorithms refer to principal component analysis (PCA), factor analysis (FA), clustering analysis (CA), positive matrix fractionation (PMF), etc., while the supervised algorithms refer to linear regression, logistic regression, support vector machine (SVM), decision tree (DT), random forest (RF), artificial neural network (ANN), and discriminant analysis (DA).

While the target parameter can be labeled, a supervised machine learning algorithm should be used in priority as accurate and stable models are expected. In the USA, the research tried to identify the source of salt ions (Mg, CL, and Na). As the samples were collected from known sites or environments, including (oceans, atmospheric deposition, weathering of common rocks, minerals and soils, and salt deposits and brines landfills, wastewater and water treatment, agriculture), the samples can be labeled. Therefore, discriminant analysis and clustering analysis were applied [35]. In Belgium, a Bayesian isotope mixing model was used to estimate proportional contributions of multiple nitrate sources in surface water [36]. In a coal mine, water inrush constantly threatens the production and human health and causes financial losses. The source apportionment technology is used in coal mines to determine the source of water inrush [37]. The water inrushes could be categized into four sources: quaternary sand-gravel pore aquifer, Dyas sandstone aquifer, limestone aquifer from Ordovician and Carboniferous, and abandoned coal mine districts, respectively. Different sources show various features and need suitable treating strategies. To set up the discriminant model, geochemical and data mining analytical protocol should be established. As the samples were collected from identified aquifers, a supervised machine learning method could be used. Huang et al. [37] proposed a technology system, the Piper-PCA-Bayes-LOOCV discrimination model to determine water inrush types in coal mines. The piper diagram is a geochemical technique to show the water characteristics, and abnormal samples/points were screened in this research. PCA was used to lower the dimension of the sample matrix, to make less variates standing for all the original variates. Then, the supervised ML model, Bayes DA, is used to train and implement a model for water source discriminant. LOOCV means leave-one-out cross-validation, to validate and improve the quality of the model. Wang et al. used discriminant analysis to determine water bursting sources in coal mines [38].

Comparing the supervised ML method, the unsupervised ML algorithms are used more frequently, for the samples are not always labeled. Pumure et al. [39] investigated the occurrence of selenium and arsenic in coal by the method of two-step PCA, founding that ultrasound leachable selenium concentrations were associated with 14 Å d-spacing phyllosilicate clays (chlorite, montmorillonite, and vermiculite all 2:1 layered clays), while ultrasound leachable arsenic concentrations were closely related to the concentration of illite, another 2:1 phyllosilicate clay. The PCA and PMF methods are often used to identify the source of trace elements. For example, lake sediment was analyzed [40] in southwest China using the PCA method, and it is shown that Cd/Hg/Pb/Zn and As were mainly from nonpoint anthropogenic sources, especially with the atmospheric emission from nonferrous metal smelting and coal consumption [41]. In Costa Rica, by using the method of PMF, eight important sources of PM 2.5 and PM 10 were identified. Vehicle exhaust, residual oil combustion, and fresh sea salt were the first three sources. Crustal, or dust aerosols originated, organic carbon and sulfate, secondary sulfate, secondary nitrate, and heavy fuels are the other potential sources [42]. In Pakistan, factor analysis was used to identify sources of surface soil contamination. It was found that Ni, Cr, Zn, and Cu were originated from industrial activates, and vehicular emission, and anthropogenic activities such as automobiles brought Pb, Cd, and Co; some other important contaminants, including Fe and Mn, were natural source origin [43]. In Turkey, the PCA was used to find latent factors that influence the water quality, mineral pollution, nutrient pollution, and organic pollution were identified to be the major factors.

3. Method

3.1 Site description and sampling

This study was carried out at the Xuzhou-Datun coal mine district, located at the northwest of Jiangsu province, eastern China (**Figure 1**). The area of Xuzhou city is in the plain of Huanghuai, South part of northern China. Sediment stratum covering the Archean system are Simian, Cambrian, middle-lower Ordovician, middle-upper Carboniferous, Permian, Jurassic, Cretaceous, Tertiary, and Quaternary system, from bottom to top. The hydrogeology cell selected for this study is isolated by a series of faults. This includes Sanhejian, Yaoqiao, and Longdong coal mines shown in **Figure 1**. In this area, groundwater flows from northeast to southwest.

The coal seams that are being mined are located in the Carboniferous and Permian systems, the former include Benxi and Taiyuan formations, and the latter include Shanxi and Lower-Shihezi formations, listed from the bottom to top in both systems. In Permian strata, there are mostly low sulfur content Gas coal and fat coal. The lower formation in Carboniferous has a higher content of sulfur than the upper layers. Mass percentage of sulfur in Permian Shanxi formation coal seams is around 0.83% in coal seam No.7 and 1.09% in coal seam No.9. In coal seam No.17 and No.19 in the Taiyuan formation, the average sulfur content was tested to be 1.87 and 3.49%, respectively. The two mining coal seams (No.2 and No.7) in the Permian system were included in this study; these are located in the middle Lower-Shihezi formations (No.2) and Shanxi formations (No.7). The two formations give thickness of 187–302.95 m and 81.67–136.13, respectively. White feldspar, quartz granule-sandstone, and silicon-mudstone cementation are the main minerals in the lower Shanxi formation. In addition, siltstone, siderite, carbon-mudstone, and plant-fossil clast can also be found. Gray mudstone, sand-mudstone, and sandstone

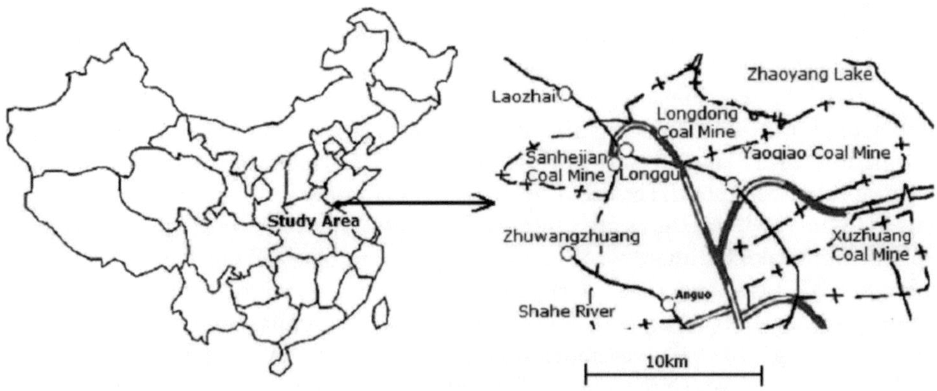

Figure 1.
Location of the study area.

are the major rocks in the middle Shanxi formation with some silicon-mudstone and siderite also present.

There are six aquifers in the sediment stratum of the hydrogeology cell. A grit aquifer in the Quaternary, a conglomerate rock aquifer in the Jurassic, two sandstone aquifers—one in the lower-Shihezi formation, and one above the coal seam in the Shanxi formation; and two limestone aquifers—one is located in the Carboniferous Taiyuan formation (thickness of 180–200 m) and the other in the Ordovician (thickness of 600 m). These last two aquifers are the main water sources of the coal seam.

3.2 Leaching experiments and sample test

A total of 16 water samples and 28 rock/coal samples were collected from the study area. Water samples were collected in 1000 mL Nalgene bottles previously acid-cleaned and rinsed twice using the water to be collected. pe and pH of water samples were taken in the field by using a JENCO 6010 pH/ORP meter. Coal and rock samples were collected from the working area at the mine and put into plastic bags that were immediately sealed.

Major ions and physical parameters of water samples were determined according to Chinese standard protocols in Jiangsu Provincial Coal Geology Research Institute. Solid samples were acid digested to determine the concentration of trace elements. The concentration of trace elements in water/coal/rock samples was determined by ICP-MS and the ICP-AES. The ICP-MS analysis was carried out in the China University of Mining and Technology using the X-Series ICP-MS—Thermo Electron Co. An internal standard of Rh was used to determine the limit of detection (0.5 pg/mL) and analytical deviation (less than 2%). The ICP-AES analysis was carried out in the Nanjing University using a JY38S ICP-AES model. The limit of detection and deviation for the analysis carried out by such equipment are 0.01 μg/mL and less than 2%, respectively.

Leaching experiments were conducted using the batch mode to simulate conditions in a coal seam where water movement is slow and dissolution reactions tend to achieve equilibrium, with regard to the previous studies [44, 45]. To simulate a "closed environment" (with low pO2; see Stumm and Morgan [46] for details), bottles were closed with a rubber stopper; samples were taken out using syringes. The pe of the solution during experiments was determined by a JENCO 6010 pH/ORP meter.

Three subsamples were used for each sample: one per 1000 mL aliquot of deionized water at the following pHs: 2, 5.6, 7, and 12. Flasks were sealed and shaken every 2 h for up to 10 days. The temperature was controlled using a water bath at about 40°C. Leachate solutions were collected using syringes at 2, 6, 24, and 48 h. A total of 0.5 mol/L HNO_3 was added into all the samples. Leachate aliquots were titrated with HCl or NaOH, depending on the pH conditions, to compare the behavior of leaching elements in acid, neutral, and alkali environments. In addition to leaching experiments, water samples including those collected from the Zhaoyang Lake and Yunlong lakes, shown in **Figure 1**, were shaken every 2 h for up to 10 days at a constant temperature of 40°C.

3.3 Multivariate analysis

While univariate statistical analysis of a large scale of data could be cumbersome and cause misunderstanding and error in the interpretation, multivariate statistical techniques are more robust. Therefore, it becomes a more useful tool for environmental data treatment and identification of anomalous patterns. During the immigration process of the trace elements from coal bedding seam to groundwater and surface water, in the complex matrix system, solid and liquid bodies are involved. In each system, the elements show different or similar coexisting patterns, and immigration behavior, including dissolution, transport, adsorption. Therefore, the multivariate analysis can be used to find out different and similar components, which suggest similar and dissimilar occurrences in solids, and immigration mechanisms during the process of water-rock interaction.

In the area of hydrochemical studies, the PCA method has been widely used to reduce dimensions and analyze the relations among the variates and samples [32–34, 47–51]. The PCA is a typical nonsupervised analytical method. To calculate the PCA result, data are first standardized by mean centering each column within the original data matrix and then dividing each of the values within each column by the column standard deviation. With PCA, the large data matrix is reduced to smaller ones that consist of PC loadings and scores. PC loadings are the eigenvectors of the correlation matrix depending on PC scores. Therefore, it contains information on all of the variables combined into a single number, with the loadings indicating the relative contribution that each variable makes to that score. PCs are calculated so that they take into account the correlations present in the original data but are uncorrelated with others. Typically, the data can be reduced to two or three dimensions representing the majority of the variance within the original data. Sometimes, more dimensions may have to be included to present more variance of the original data [33]. Based on the PCA analytical result, the loadings and scores of the data frame were then clustered in the dimensions that PCA has reduced. As the axis of coordinates was rotated to achieve maximum loadings of elements, the rotated axis of coordinates was marked as RCs.

The bi-plot of the PCA result is usually drawn to show patterns of parameters and samples. However, the loading and score of the PCA axis show different aspects of the result. In our study, the loadings of every drawn show coexisting pattern of elements, and scores of every drawn show the coexisting pattern of samples. What we focus on is the coexisting pattern of elements to disclose their migration mechanism. The clustering result of loadings shows similar and different patterns among elements and parameters. Therefore, the coexisting behavior of elements and parameters can be summarized. The clustering result of scores shows similar and different patterns among samples. Therefore, the coexisting behavior of samples, which means types of solid and liquid samples, can be summarized. The clustering method was based on the Gaussian mixture model. The GM model can cluster target reasonably. Comparing with K-means algorithm, the GM model does not divide the different group by stiff border but allows some mixture of different groups. So, the classifying probability for each group can be calculated.

We have applied software R as a tool, the packages psych and mclust were used to calculate PCA and GM model clustering results.

4. Result and discussion

4.1 Geochemical analysis

A total of 16 water samples were collected from the study site, including 12 coal mine waters, two surface waters, and two carbonate waters, respectively. Concentrations of major ions are drawn in a piper plot (**Figure 2**). **Figure 2** suggests that the carbonate water and coal mine water belong to medium-mineralized water, and surface water belongs to low-mineralized water, respectively. The surface water is Na-Mg-Ca-Cl^--SO_4^{2-}-HCO_3^--type water, the carbonate water is

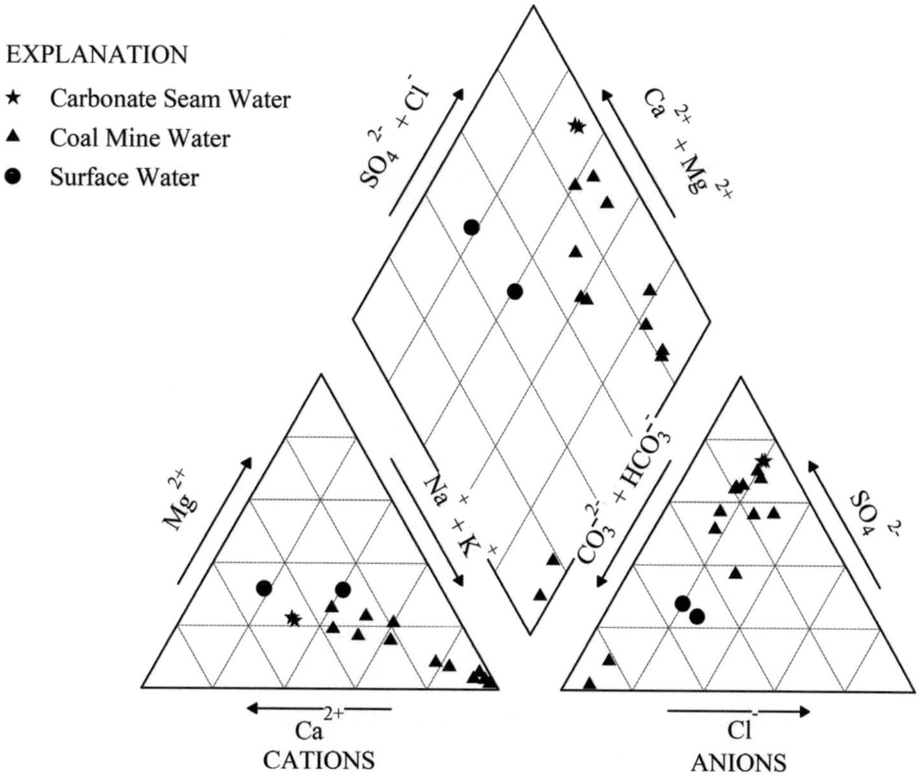

Figure 2.
Piper plot of the water samples.

Na-Mg-Ca-SO_4^{2-}-type water, and the coal mine water is Na-Ca-SO_4^{2-}-, Na-SO_4^{2-}-, or Na-HCO_3^--type water, respectively. Coal mine waters showed characteristics of high-soluble minerals. [SO_4^{2-}] of most coal mine water samples were higher than USEPA and Chinese highest limit, 250 mg/L. Besides [SO_4^{2-}], [Cl^-], TDS, and hardness were also higher than the Chinese-regulated limit. The combination of higher levels of Ca^{2+}, Mg^{2+}, HCO_3^-, and SO_4^{2-} concentrations in the groundwater suggests that the coupled reactions involving sulfide oxidation and carbonate dissolution largely control the solute acquisition processes in the study area [52].

The PCA analysis is used to reduce the dimensions of the water matrix. In this study case, dimension means water parameters. Water samples are represented by 10s of conventional inorganic and organic parameters, some of which are an indicator of the environment and reaction pathways, and some others a redundant or collinear. The PCA method could solve problems of not only parameter redundant and collinear, but also shows principal components in the data matrix, and relationships between parameters and among the parameters and samples could also be shown by using the parameters' loading and samples' score, respectively.

In this study, the traditional method of PCA calculation was applied, and principal components and variance that the PC explained were calculated. In the original table, 16 parameters were tested, and the PCA calculation used 16 new components to represent the original parameters, which explain the variance of samples, in descending order. The head six components explained 29, 21, 17, 10, 9, and 5% of the variance, respectively. Considering the balance of more variance explained and less components, we chose two principal components to stand for the sample data. The GM method was used to group the ions and trace elements in the water sample, which is shown in **Figure 3**. The parameters were clustered into four

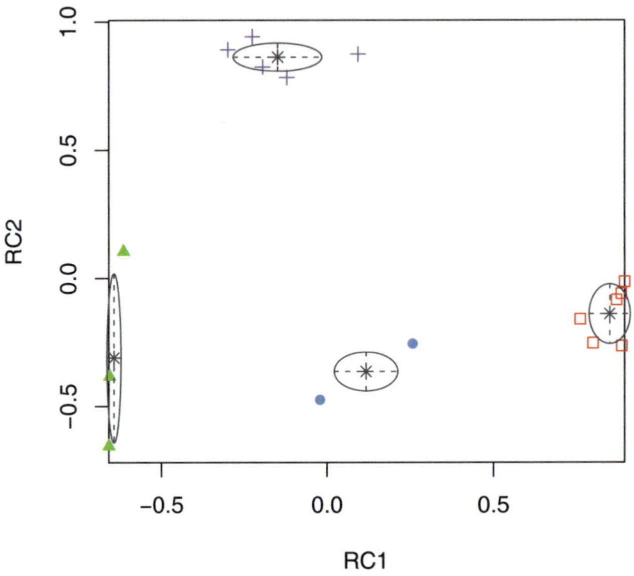

Figure 3.
Loadings of the multivariate analysis and clustering result of water samples.

groups. Group 1 includes K^+ + Na^+ and Cl^-; group 2 includes Ca^{2+}, Mg^{2+}, Cl^-, SO_4^{2-}, TDS, and hardness; group 3 includes HCO_3^-, CO_3^{2-}, and pH; group 4 includes As, Hg, Se, Cd, Pb, respectively. The samples were collected in or around the coal mine district, so the clustering result is representative, and the groups were separated from others distinctly. From the clustering result, it is suggested that group 2 stands for the dissolution of carbonate, and group 4 stands for the trace element. The trace element contaminant could be identified from this result.

4.2 Leaching mechanism of trace elements from the coal host rock

To investigate the leaching mechanism of trace elements from the coal host rock, both the rock sample and water sample were tested. The rock samples were those collected from coal roof, which then was processed in a standard treatment to decide its content. The milled rock samples were mixed with deionized water in the batch experiments to observe and evaluate the leaching behavior and mechanism of the trace elements from rock to water. The major and trace element concentrations in host rock and leachate are listed in the Table 1 in Shan et al. [53]. A hypothesis was that the occurrence and leaching mechanism of the trace elements in the solid samples were related to their concentrations in the water samples. Therefore, the PCA was applied to reduce dimensions of the rock and water samples, and then, the analytical results of solid and liquid samples are discussed parallelly.

For the rock samples, 18 elements were tested, and then, the PCA method was applied. The first two components explained 91% of all variance; therefore, the two PCs were used to stand for information of the data. For the water samples, 16 ions and trace elements were tested. The same analytical process was applied. The first two PCs explained 87% of all variance, which were used to stand for information in the water samples. By using the new PCs, parameters were assigned loadings on every new component. Then, the parameters of rock and water samples can be drawn in a two-dimensional (2D) scatter diagram. **Figure 4** shows the elements of rock samples, and **Figure 5** shows the ions and elements of water samples in a 2D scatter diagram, respectively.

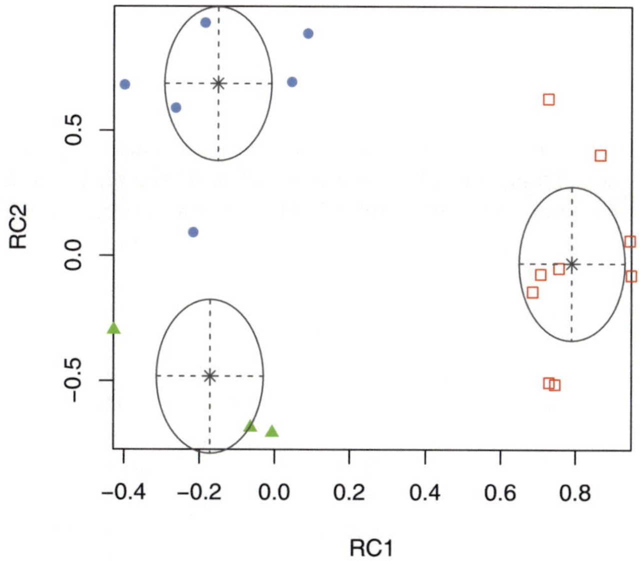

Figure 4.
Loadings of the multivariate analysis and clustering result of rock samples.

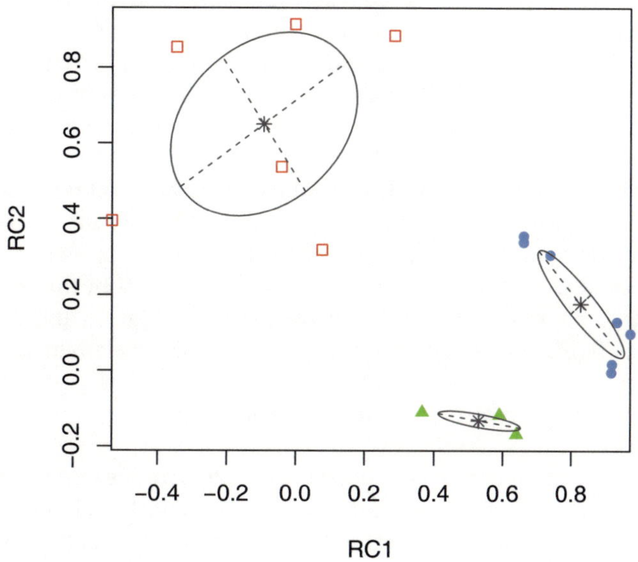

Figure 5.
Loadings of the multivariate analysis and clustering result of rock leachate.

The PCA-treated data were clustered using the expectation maximization (EM) algorithm. The EM algorithm could make several clustering results. By considering the BIC score and conciseness of every clustering model, the parameters in the rock samples were clustered into three groups. The first group includes Mo, Pb, Cr, V, Ti, and Al, which are marked in solid circles; the second group includes Zn, Ba, Mn, Fe, Mg, As, Hg, Se, and Cd, which are shown in hollow squares; the third group includes Cu, Sr, and Ca, which are shown in solid triangles. As mentioned before, the clustering could help to analyze the elements' occurrence in solid samples. Cr has a high affinity of clay and ash yield in gangue [3]. Zhou et al. [2] reported a high relationship of Pb and Se and with Fe in gangue, so high-sulfide mineral affinity was observed. Zn and Cd were found to have a high association with pyrite and sphalerite. Xiong et al. [26] found that Cd is mainly in sulfide form in the coal host rock. As and Mo are mainly carbonate- and silicate-related form. Finkelman et al. [3] found that Mo, Pb, Cr, Ti, and Al are mainly in clay minerals, As, Hg, Cd, and Zn mainly occur in sulfide form, and Ca and Sr are mainly carbonate-related. The PCA analysis corroborates the previous studies. As the **Figure 5** shows, the first group stands for clay affinity elements, the second group stands for elements with sulfur-mineral affinity, and the third group stands for the carbonate-related elements.

The ions and trace elements in the rock leachate could be clustered into three groups, the first group includes Al, Si, Cr, Mn, Fe, Cd, and Pb; the second group includes Ti, V, As, Se, Mo, and Hg; and the third group includes Zn, Sr, and Ba, respectively. The coexisting pattern of ions and elements in the water are controlled not only the occurrence in rock, but also the water-rock interaction, and adsorption behavior. Therefore, the clustering result of solid and liquid results was not exactly the same. However, two results are comparable to find out certain or probable reaction mechanisms in the water-rock interaction pathway. The three groups clustered for the water samples can be compared with those of the solid samples. Therefore, a primary deduction could be made. The first group of elements in the water samples suggests the reaction pathway of clay reaction with water. When the clay mineral reacts with water, the transformation of illite to kaolinite could happen, and some minerals, such as Cr, could be released. Cd was clustered to the second group in the rock analysis but was clustered to group 1 in water analysis. The result could be explained by two reasons: first, Pb and Cd embedded in both sulfur minerals and clay minerals, and second, Pb and Cd were controlled not only by dissolution, but also by adsorption. When the water has a low pH value, metal elements tend to release, while they could be adsorbed in a higher pH environment. According to our observation, the concentration of Pb and Cd in the surface water in the coal mine district was evidently higher than that in the non-coal mine district. As, Hg, and Se have a similar pattern in the solid and liquid samples. It is apparent that they were controlled by the dissolution of sulfur minerals. The content of the sulfur mineral in the rock was not high in our samples. However, the oxidation and dissolution processes were distinct, leading to the release of toxic trace elements.

4.3 Leaching mechanism of trace elements from coal

The major and trace element concentrations in coal and leachate are listed in the Table 1 in Shan et al. [53]. The same analytical method with rock was applied to the coal and coal leaching analysis. And the PCA and clustering analytical results of coal and coal leaching water are shown in **Figures 6** and 7. Two principal components could explain 96 and 91% variance for the coal and leachate, respectively. As **Figure 6** shows that elements are clustered into four groups, the group 1 includes Mo, Pb, Cr, V, Cu, Ti, Al, Hg, and Se; group 2 includes Zn and Cd; group 3 includes Ba, Mn, Sr, Mg, and Ca; group 4 includes Fe and As, respectively. The ions and trace elements in coal leachate, as shown in **Figure** 7, were grouped into three groups. Group 1 includes Al, Se, and Pb; group 2 includes Si, As, Sr, Mo, and Hg; group 3 includes Ti, Cr, Mn, Fe, Zn, Cd, and Ba, respectively. Finkelman et al. [3] investigated the occurrence of most of the trace elements, it is found that 65% of Ti, 90% of Al, and 75% of Cr 25% and 30% of Cu and Mo are in clay minerals, little Pb and Se are in clay form, 75 and 65% of Zn and Cd formed in mono-sulfide form, and 70 and 90% of As and Hg are sulfide form. Pumure et al. [39] argued that As and Se usually occur in clay minerals. Pb was found to be sulfide form as pyrite and galena [54] and organic form [55].

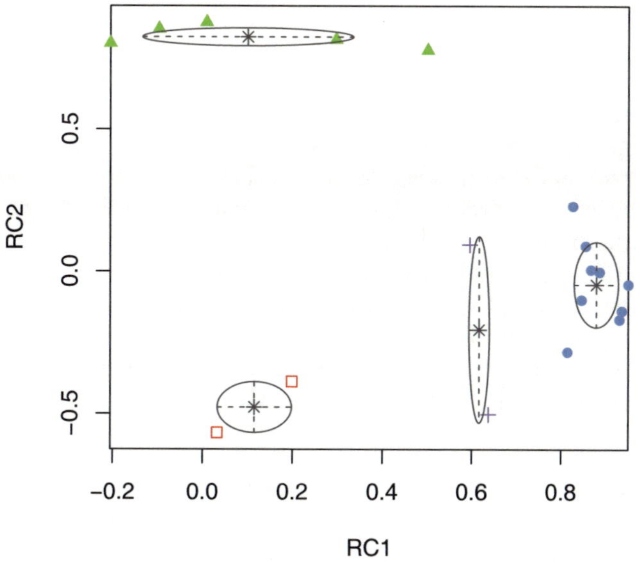

Figure 6.
Loadings of the multivariate analysis and clustering result of coal samples.

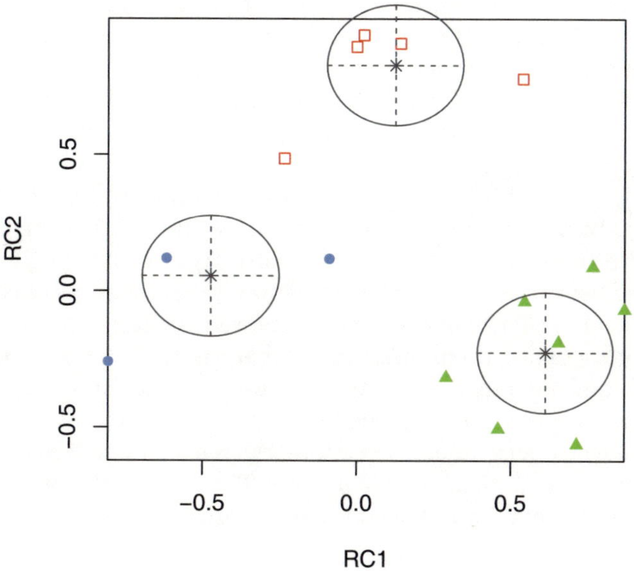

Figure 7.
Loadings of the multivariate analysis and clustering result of coal leachate.

Combining the literature review and PCA-clustering analysis, group 1 for the coal samples stands for clay affinity, groups 2 and 4 are sulfur-mineral elements, and group 3 is related to carbonate minerals. Group 2 has two elements, Zn and Cd. This result is consistent with some previous studies [2, 56]. It is concluded the main occurrence of trace elements: As, Hg, Cd occurred in sulfide minerals, and Pb, Cr, and Se occurred in clay minerals, respectively. Zn and Cd are the primary elements in sphalerite. Compared with the host rock, the sphalerite is more probably to form an independent mineral in coal.

The coal leachate clustering results were relatively different with that of the analytical results of coal. Compared to the rock samples, coal is a more complex matrix and consists of organic and mineral matter, the latter including crystalline minerals, non-crystalline mineraloids, and elements with non-mineral associations [55]. However, some patterns could be concluded. Group 1 includes Al, Se, and Pb, which is similar to group 1 in the coal analysis. Therefore, group 1 stands for the elements that originated from clay minerals. Group 2 stands for the elements related to sulfur-bearing minerals. As and Hg had similar behavior patterns in solid and liquid matrices. So the leaching product in water was mainly from the dissolution of its bearing mineral, the sulfide mineral. Similar to the host rock analysis, low content of sulfur-mineral may lead to trace element concentration.

The trace elements Se, Cr, and Pb have similar behavior patterns in solid and liquid matrices, suggesting a dissolution progress of its bearing minerals. According to the literature research and coexisting analysis, these elements usually occur in continental facies minerals, such as clay minerals.

5. Conclusion

A data mining workflow, composed of principal component analysis and the Gaussian mixture model, was applied to find the trace elements' occurrence and leaching mechanism from coal and rock to surface and groundwater bodies. It is found that Se, Cd, Hg, and As were associated with sulfide minerals; Be and V occurred in carbonate minerals; Cr and Pb occurred mainly in clay minerals in the rock samples. While As and Hg were mainly occurred in sulfide minerals, Se, Cr, and Pb were embedded in clay minerals.

When the host rock is leaching with water, As, Hg, and Se were originated from oxidation and dissolution of sulfur-mineral; especially for pyrite, Cr was mainly controlled by the transformation of clay minerals. When the coal is leaching with water, As and Hg showed high affinity of sulfur-minerals, and Se and Cr seemed to be controlled by the water-rock interaction of clay minerals. It suggested that Se exist in sulfide mineral, clay minerals, and also organic matters. Therefore, the leaching mechanism of Se is not unique, and multiple mechanisms may control or influence the leaching behaviors. Cd and Pb showed apparent differences between the solid samples and liquid samples. The mechanism leading to this result was probably explained not only the releasing process, but also the adsorption process. These elements are typical metal elements. They can be easily adsorbed in the alkaline and neutral environment. Therefore, the released metal elements were adsorbed by clay minerals and organic matters. The immigration mechanism and long-term environmental impact need further studies.

Acknowledgements

The test of samples was carried out in the Jiangsu Provincial Coal Geology Research Institute, the Analysis and Test Center of the China University of Mining and Technology, Imperial College London. We would like to thank all of them for their support.

Author details

Yao Shan
School of Emergency Technology and Management, North China Institute of Science and Technology, Yanjiao, China

*Address all correspondence to: 9106350@qq.com

References

[1] Goodell J. Big Coal: The Dirty Secret Behind America's Energy Future. Houghton-Mifflin: New York, NY; 2006

[2] Zhou C, Liu G, Fang T, Sun R, Wu D. Leaching characteristic and environmental implication of rejection rocks from Huainan Coalfield, Anhui Province, China. Journal of Geochemical Exploration. 2014;**143**:54-61

[3] Finkelman RB, Plamer CA, Wang P. Quantification of the modes of occurrence of 42 elements in coal. International Journal of Coal Geology. 2018;**185**:138-160

[4] Fang WX, Wu PW, Hu RZ. Geochemical research of the impact of Se–Cu–Mo–V-bearing coal layers on the environment in Pingli County, Shaanxi Province, China. Journal of Geochemical Exploration. 2003;**80**:105-115

[5] Finkelman RB, Orem W, Castranova V, Tatu CA, Belin HE, Zheng B, et al. Health impacts of coal and coal use: Possible solutions. International Journal of Coal Geology. 2002;**50**:425-443

[6] Liu G, Yang P, Peng Z, Chou CL. Petrographic and geochemical contrasts and environmentally significant trace elements in marine-influenced coal seams, Yanzhou mining area, China. Journal of Asian Earth Sciences. 2004;**23**:491-506

[7] Liu G, Vassilev SV, Gao L, Zheng L, Peng Z. Mineral and chemical composition and some trace element contents in coals and coal ashes from Huaibei coal field, China. Energy Conversion and Management. 2005;**46**:2001-2009

[8] Querol X, Alastuey A, Zhuang X, Hower JC, Lopez-Soler A, Plana F, et al. Petrology, mineralogy and geochemistry of the Permian and Triassic coals in the Leping area, Jiangxi Province, southeast China. International Journal of Coal Geology. 2001;**48**:23-45

[9] Mohanty AK, Lingaswamy M, Rao G, Sankaran S. Impact of acid mine drainage and hydrogeochemical studies in a part of Rajrappa coal mining area of Ramgarh District, Jharkhand State of India. Groundwater for Sustainable Development. 2018;**7**:164-175

[10] Sahoo PK, Tripathy S, Panigrahi MK, Equeenuddin SM. Geochemical characterization of coal and waste rocks from a high sulfur bearing coalfield, India: Implication for acid and metal generation. Journal of Geochemical Exploration. 2014;**145**:135-147

[11] Zhu C, Qu S, Zhang J, Wang Y, Zhang Y. Distribution, occurrence and leaching dynamic behavior of sodium in Zhundong coal. Fuel. 2017;**190**:189-197

[12] Zhao F, Sun H, Liu N, Cai W, Han R, Chen B. Evaluation of static acid production potential for coal bearing formation (in Chinese). Earth Science-Journal of China University of Geosciences. 2014;**39**(3):350-356

[13] Cravotta CA III. Monitoring, field experiments, and geochemical modeling of Fe (II) oxidation kinetics in a stream dominated by net-alkaline coal-mine drainage, Pennsylvania, USA. Applied Geochemistry. 2015;**62**:96-107

[14] Cravotta CA III, Brady KBC. Priority pollutants and associated constituents in untreated and treated discharges from coal mining or processing facilities in Pennsylvania, USA. Applied Geochemistry. 2015;**62**:108-130

[15] Bidari E, Aghazadeh V. Pyrite oxidation in the presence of calcite and dolomite: Alkaline leaching, chemical

modeling and surface characterization. Transactions of the Nonferrous Metals Society of China. 2018;**28**:1433-1443

[16] Boruvka L, Kozak J, Muhlhanselova M, Donatova H, Nikodem A, Nemecek K, et al. Effect of covering with natural topsoil as a reclamation measure on brown- coal mining dumpsites. Journal of Geochemical Exploration. 2012;**113**:118-123

[17] Dai S, Dan L, Chou CL, Zhao L, Zhang Y, Ren D, et al. Mineralogy and geochemistry of boehmite-rich coals: New insights from the Haerwusu Surface Mine, Jungar Coalfield, Inner Mongolia, China. International Journal of Coal Geology. 2008;**74**:185-202

[18] Paul D. Petrology and geochemistry of the Salma dike, Raniganj coalfield (Lower Gondwana), eastern India: Linkage with Rajmahal or Deccan volcanic activity? Journal of Asian Earth Sciences. 2005;**25**:903-913

[19] Zheng G, Liu G, Chou CL, Qi C, Zhang Y. Geochemistry of rare earth elements in Permian coals from the Huaibei Coalfield, China. Journal of Asian Earth Sciences. 2007;**31**:167-176

[20] Karayiğit AI, Bircan C, Mastalerz M, Oskay RG, Querol X, Lieberman NR. Ibrahim Türkmen Coal characteristics, elemental composition and modes of occurrence of some elements in the İsaalan coal (Balıkesir, NW Turkey). International Journal of Coal Geolog. 2017;**172**:43-59

[21] Liu J, Zong Y, Yan X, Ji D, Yang Y, Hu L. Modes of occurrence of highly-elevated trace elements in superhigh-organic-sulfur coals. Fuel. 2015, 2015;**156**:190-197

[22] Tozsin G. Hazardous elements in soil and coal from the Oltu coal mine district, Turkey. International Journal of Coal Geology. 2014;**131**:1-6

[23] Song D, Qin Y, Zhang J, Wang W, Zheng C. Concentration and distribution of trace element in some coals from Northern China. International Journal of Coal Geology. 2007;**69**:179-191

[24] Zheng L, Liu G, Wang L, Chou CL. Composition and quality of coals in the Huaibei Coalfield, Anhui, China. Journal of Geochemical Exploration. 2008;**97**:59-68

[25] Sia S, Abdullah WH. Concentration and association of minor and trace elements in Mukah coal from Sarawak, Malaysia, with emphasis on the potentially hazardous trace elements. International Journal of Coal Geology. 2011;**88**:179-193

[26] Xiong Y, Xiao T, Liu Y, Zhu J, Ning Z, Xiao Q. Occurrence and mobility of toxic elements in coals from endemic fluorosis areas in the Three Gorges Region, SW China. Ecotoxicology and Environmental Safety. 2017;**144**:1-10

[27] Wang W, Hao W, Bian Z, Lei S, Wang X, Sang S, et al. Effect of coal mining activities on the environment of Tetraena mongolica in Wuhai, Inner Mongolia, China—A geochemical perspective. International Journal of Coal Geology. 2014;**132**:94-102

[28] Dai S, Li D, Ren D, Tang Y, Shao L, Song H. Geochemistry of the late Permian No. 30 coal seam, Zhijin Coalfield of Southwest China: Influence of a siliceous low-temperature hydrothermal fluid. Appl. Geochemistry. 2004;**19**:1315-1330

[29] Orakwe LC, Chukwuma EC. Multivariate analysis of ground water characteristics of Ajali sandstone formation: A case study of Udi and Nsukka LGAs of Enugu State of Nigeria.

Journal of African Earth Sciences. 2017;**129**:668-674

[30] Matiatos I, Paraskevopoulos V, Lazogiannis K, Botsou F, Dassenakis M, Ghionis G, et al. Surface-ground water interactions and hydrogeochemical evolution in a fluvio-deltaic setting: The case study of the Pinios River delta. Journal of Hydrology. 2018;**561**:236-249

[31] Zhu B, Wang X, Rioual P. Multivariate indications between environment and ground water recharge in a sedimentary drainage basin in northwestern China. Journal of Hydrology. 2017;**549**:92-113

[32] Hwang CK, Cha JM, Kim KW, Lee HK. Application of multivariate statistical analysis and ageographic information system to trace element contamination in the Chungnam Coal Mine area, Korea. Applied Geochemistry. 2001;**16**:1455-1464

[33] Singh KP, Malik A, Mohan D, Sinha S. Multivariate statistical techniques for the evaluation of spatial and temporal variations in water quality of Gomti River (India)—A case study. Water Research. 2004;**38**(18):3980-3992

[34] Liu P, Hoth N, Drebenstedt C, Sun Y, Xu Z. Hydro-geochemical paths of multi-layer groundwater system in coal mining regions—Using multivariate statistics and geochemical modeling approaches. Science of the Total Environment. 2017; **601-602**:1-14

[35] Hajigholizadeh M, Melesse AM. Assortment and spatiotemporal analysis of surface water quality using cluster and discriminant analyses. Catena. 2017;**151**:247-258

[36] Xue D, De Baets B, Van Cleemput O, Hennessy C, Berglund M, Boeckx P. Use of a Bayesian isotope mixing model to estimate proportional contributions of multiple nitrate sources in surface water. Environmental Pollution. 2012;**161**:43-49

[37] Huang P, Yang Z, Wang X, Ding F. Research on Piper-PCA-Bayes-LOOCV discrimination model of water inrush source in mines. Arabian Journal of Geosciences. 2019;**12**(334):1-14.

[38] Wang J, Li X, Cui T, Yang J. Application of distance discriminant analysis method to headstream recognition of water-bursting source. Procedia Engineering. 2011;**26**:374-381

[39] Pumure I, Renton JJ, Smart RB. The interstitial location of selenium and arsenic in rocks associated with coal mining using ultrasound extractions and principal component analysis (PCA). Journal of Hazardous Materials. 2011;**198**:151-158

[40] Lin Q, Liu E, Zhang E, Li K, Shen J. Spatial distribution, contamination and ecological risk assessment of heavy metals in surface sediments of Erhai Lake, a large eutrophic plateau lake in southwest China. Catena. 2016; **145**:193-203

[41] Tian HZ, Zhu CY, Gao JJ, Cheng K, Hao JM, Wang K, et al. Quantitative assessment of atmospheric emissions of toxic heavy metals from anthropogenic sources in China: Historical trend, spatial distribution, uncertainties, and control policies. Atmospheric Chemistry and Physics. 2015;**15**(17):10127-10147

[42] Murillo JH, Roman SR, Rojas Marin JF, Ramos AC, Jimenez SB, Gonzalez BC, et al. Chemical characterization and source apportionment of PM10 and PM2.5 in the metropolitan area of Costa Rica, Central America. Atmospheric Pollution Research. 2013;**4**(2):181-190

[43] Malik RN, Jadoon WA, Husain SZ. Metal contamination of surface soils of industrial city Sialkot, Pakistan: A multivariate and GIS approach.

Environmental Geochemistry and Health. 2010;**32**(3):179-191

[44] Su T, Wang J. Modeling batch leaching behavior of arsenic and selenium from bituminous coal fly ashes. Chemosphere. 2011;**85**:1368-1374

[45] Schwartz GE, Rivera N, et al. Leaching potential and redox transformations of arsenic and selenium in sediment microcosms with fly ash. Applied Geochemistry. 2016;**67**:177-185

[46] Stumm W, Morgan JJ. Aquatic Chemistry: An Introduction Emphasizing Chemical Equilibria in Natural Waters. University of Michigan: Wiley-Interscience Publications; 1981. p. 780

[47] Güler C, Kurt MA, Alpaslan M, Akbulut C. Assessment of the impact of anthro-pogenic activities on the groundwater hydrology and chemistry in Tarsus coastal plain (Mersin, SE Turkey) using fuzzy clustering, multivariate statistics and GIS techniques. Journal of Hydrology. 2012;**414-415**:435-451

[48] Sako A, Bamba O, Gordio A. Hydrogeochemical processes controlling groundwa- ter quality around Bomboré gold mineralized zone, Central Burkina Faso. Journal of Geochemical Exploration. 2016;**170**:58-71

[49] Cortes JE, Muñoz LF, Gonzalez CA, Niño JE, Polo A, Suspes A, et al. Hydrogeochemistry of the formation waters in the San Francisco field, UMV basin, Colombia—A multivariate statistical approach. Journal of Hydrology. 2016;**539**:113-124

[50] Carucci V, Petitta M, Aravena R. Applied Geochemistry Interaction between shallow and deep aquifers in the Tivoli Plain (Central Italy) enhanced by groundwater extraction: A multi-isotope approach and geochemical modeling. Applied Geochemistry. 2012;**27**(1):266-280

[51] Chihi H, de Marsily G, Belayouni H, Yahyaoui H. Relationship between tectonic structures and hydrogeochemical compartmentalization in aquifers: Example of the "Jeffara de Medenine" system, south-east Tunisia. Journal of Hydrology. 2015;**4**:410-430

[52] Singh AK, Mondal GC, Singh S, Singh PK, Singh TB, Tewary BK, et al. Aquatic geochemistry of Dhanbad district, coal city of India: Source evaluation and quality assessment. Journal of the Geological Society of India. 2007;**69**:1088-1102

[53] Shan Y, Wang W, Qin Y, Gao L. Data on trace element concentrations in coal and host rock and leaching product in different pH values and open/closed environments. Data in Brief. 2019;**25**:1-13

[54] Dai S, Ren D, Chou CL, Li S, Jiang Y. Mineralogy and geochemistry of the No. 6 coal (Pennsylvanian) in the Junger Coalfield, Ordos Basin, China. International Journal of Coal Geolog. 2006;**66**:253-270

[55] Dai S, Hower JC, Finkelman RB, Graham IT, French D, Ward CR, et al. Organic associations of non-mineral elements in coal: A review. International Journal of Coal Geology. 2020;**218**:103347

[56] Gurdal G. Geochemistry of trace elements in Can coal (Miocene), Canakkale, Turkey. International Journal of Coal Geolog. 2008;**74**:28-40

Chapter 11

Tourist Sentiment Mining Based on Deep Learning

Weijun Li, Qun Yang and Wencai Du

Abstract

Mining the sentiment of the user on the internet via the context plays a significant role in uncovering the human emotion and in determining the exactness of the underlying emotion in the context. An increasingly enormous number of user-generated content (UGC) in social media and online travel platforms lead to development of data-driven sentiment analysis (SA), and most extant SA in the domain of tourism is conducted using document-based SA (DBSA). However, DBSA cannot be used to examine what specific aspects need to be improved or disclose the unknown dimensions that affect the overall sentiment like aspect-based SA (ABSA). ABSA requires accurate identification of the aspects and sentiment orientation in the UGC. In this book chapter, we illustrate the contribution of data mining based on deep learning in sentiment and emotion detection.

Keywords: Deep learning, Aspect-based Sentiment Analysis, User-generated content, Gated Recurrent Neural Network

1. Introduction

Since the world has been inundated with the increasing amount of tourist data, tourism organizations and business should keep abreast about tourist experience and views about the business, product and service. Gaining insights into these fields can facilitate the development of the robust strategy that can enhance tourist experience and further boost tourist loyalty and recommendations. Traditionally, business rely on the structured quantitative approach, for example, rating tourist satisfaction level based on the Likert Scale. Although this approach is effective to prove or disprove existing hypothesis, the closed ended questions cannot reveal exact tourist experience and feelings of the products or services, which hampers obtaining insights from tourists. Actually, business have already applied sophisticated and advanced approaches, such as text mining and sentiment analysis, to disclose the patterns hidden behind the data and the main themes.

Sentiment analysis (SA) has been used to deal with the unstructured data in the domain of tourism, such as texts, images, and video to investigate decision-making process [1], service quality [2], destination image and reputation [3]. As for the level of sentiment analysis, it has been found that most extant sentiment analysis in the domain of tourism is conducted at document level [4–7]. Document-based sentiment analysis (DBSA) regards the individual whole review or each sentence as an independent unit and assume there is only one topic in the review or in the sentence. However, this assumption is invalid as people normally express their semantic orientation on different aspects in a review or a sentence [8]. For example, in the sentence "we had impressive breakfast, comfortable bed and friendly and professional staff serving us", the aspects discussed here are "breakfast", "bed" and "staff" and the users give positive comments on these aspects ("impressive", "comfortable" and "friendly and professional"). Since the sentiment obtained through DBSA is at coarse level, aspect-based sentiment analysis (ABSA) has been suggested to capture sentiment tendency of finer granularity.

To obtain the sentiment at the finer level, ABSA has been proposed and developed over the years. ABSA normally involves three tasks, the extraction of opinion target (also known as the "aspect term"), the detection of aspect category and the classification of sentiment polarity. Traditional methods to extract aspects rely on the word frequency or the linguistic patterns. Nevertheless, it cannot identify infrequent aspects and heavily depends on the grammatical accuracy to manipulate the rules [9]. As for the detection of sentiment polarity, supervised machine learning approaches, like Maximum Entropy (ME), Conditional Random Field (CRF) and Support Vector Machine (SVM). Although machine learning-based approaches have achieved desirable accuracy and precision, they require huge dataset and manual training data. In addition, the results cannot be duplicated in other fields [10]. To overcome these shortcomings, ABSA of deep learning (DL) approaches has the advantage of automatically extracting features from data [9]. Extant studies based on DL methods in tourism have investigated and explored tourist experiences in economy hotel [11], the identification of destination image [12], review classification [13]. Although DL methods have been applied in tourism, ABSA in tourism is scant. Therefore, this study reviewed sentiment analysis at aspect level conducted by DL approaches, compared the performance of DL models, and explored the model training process.

With the references of surveys about DL methods [9, 14], this study followed the framework of ABSA proposed by Liu (2011) [8] to achieve the following aims: (1) provide an overview of the studies using DL-based ABSA in tourism for researchers and practitioners; (2) provide practical guidelines including data annotation, pre-processing, as well as model training for potential application of ABSA in similar areas; (3) train the model to classify sentiments with the state-of-art DL methods and optimizers using datasets collected from TripAdvisor. This paper is organized as follows: Section 2 reviews the cutting-edge techniques for ABSA, studies using DL for NLP tasks in tourism, and research gap; Section 3 presents the annotation schema of the given corpus and DL methods used in this study; Section 4 describes the details of annotation results, model training, and the experiment results. Section 5 provides the conclusions and future extensions.

2. Literature review

An extensive literature review of the state-of-art techniques for ABSA and the studies using DL in tourism is provided in this section.

2.1 Input vectors

To convert the NLP problems into the form that computers can deal with, the texts are required to be transformed into a numerical value. In ML-based approaches, One-hot and Counter Vectorizer are commonly used. One-hot encoding can realize a token-level representation of a sentence. However, the use of One-hot encoding usually results in high dimension issues, which is not computationally efficient [15]. Another issue is the difficulty of extracting meanings as this approach assumes that words in the sentence are independent, and the similarities cannot be measured by distance nor cosine-similarity. As for Counter Vectorizer, although it can convert the whole sentence into one vector, it cannot consider the sequence of the words and the context.

Nevertheless, in DL based approaches, pre-trained word embeddings have been proposed in [16, 17]. Word embedding, or word representation, refers to the learned representation of texts in which the words with identical meanings would have similar representation. It has been proved that the use of word embeddings as the input vectors can make a 6–9% increase in aspect extraction [18] and 2% in the identification of sentiment polarity [19]. Pre-trained word embeddings are favored as random initialization could result in stochastic gradient descent (SGD) in local minima [20]. Based on the network language model, a feedforward architecture, which combined a linear projection layer and a non-linear hidden layer, could learn the word vector representation and a statistical language model [21].

Word2Vec [16] proposed the skip-gram and continuous bag-of-words (CBOW) models. By setting the window size, skip-gram can predict the context based on the given words, while the CBOW can predict the word based on the context. Frequent words also are assigned binary codes in Huffman trees because Also, due to the fact that the word frequency is appropriate to acquire classes in neural net language models, frequent words are assigned binary codes in Huffman trees. This practice in Word2Vec helps reduce the number of output units that are required to be assessed. However, the window-based approaches of Word2Vec do not work on the co-occurrence of the text and do not harness the huge amount of repetition in the texts. Therefore, to capture the global representation of the words in all sentences, GloVe can take advantage of the nonzero elements in a word-word cooccurrence matrix [17]. Although the models discussed above performed well in similarity tasks and named entity recognition, they cannot cope with the polysemous words. In a more recent development, Embeddings from language model (ELMo) [22], Bi-directional Encoder Representations from Transformers (BERT) [23] can identify the context-sensitive features in the corpus. The main difference between these two architectures is that ELMo is feature-based, while BERT is deeply bidirectional. To be specific, the contextual representation of each token is obtained through the concatenation of the left-to-right and right-to-left representations. In contrast, BERT applies masked language models (MLM) to acquire the pre-trained deep bidirectional representations. MLM can randomly mask certain tokens from the input and predict the ID of the input depending only on the context. Additionally, BERT is capable of addressing the issues of long text dependence.

Nonetheless, researchers have combined certain features with word embedding to produce more pertinent results. These features include Part-Of-Speech (POS) and chunk tags, and commonsense knowledge. It has been observed that aspect terms are usually nouns or noun phrases [8]. The original word embeddings of the texts are concatenated with as k-dimensional binary vectors that represent the k POS, or k tags. The concatenated word embeddings are fed into the models (Do et al.,, Prasad, Maag, and Alsadoon, 2019 [9]). It has been proved that the use of POS tagging as input can improve the performance of aspect extraction, with gains from 1% [18, 20] to 4% [24]. Apart from the POS, concepts that are closely related to the affections are suggested to be added as word embeddings [25, 26]. POS focused on the grammatical tagging of the words in a corpus, while concepts that are extracted from SenticNet emphasize the multi-word expressions and the dependency relation between clauses. For example, the multi-word expression "win lottery" could be related to the emotions "Arise-joy" and the single-word expression "dog" is associated with the property "Isa-pet" and the emotions "Arise-joy" [26]. After being parsed by SenticNet, the obtained concept-level information (property and the emotions) is embedded into the deep neural sequential models. The performance of the Long Short-Term Memory (LSTM) [27] combined with SenticNet exceeded the baseline LSTM [26].

2.2 DL methods for ABSA

This section reviews the DL methods used for ABSA, including Convolutional Neural Network (CNN), Recurrent Neural Network (RNN), Attention-based RNN, and Memory Network.

2.2.1 CNN

CNN can learn to capture the fixed-length expressions based on the assumption that keywords usually include the aspect terms with few connections of the positions [28]. Besides, as CNN is a non-linear model, it usually outperforms the linear-model and rarely relies on language rules [29]. A local feature window of 5 words was firstly created for each word in the sentence to extract the aspects. Then, a seven-layer of CNN was tested and generated better results [29]. To capture the multi-word expressions, the model proposed [30] contained two separate convolutional layers with non-linear gates. N-gram features can be obtained by the convolutional layers with multiple filters. Li et al. [13] put position information between the aspect words and the context words into the input layer in CNN and introduced the aspect-aware transformation parts. Fan et al. [31] integrated the attention mechanism with a convolutional memory network. This proposed model can learn multi-word expressions in the sentence and identify long-distance dependency.

Apart from simply extracting the aspects alone, CNN can identify the sentiment polarity at the same time, which can be regarded as multi-label tasking classification or multitasking issues. As for researchers who considered ABSA multi-label tasking classification, a probability distribution threshold was applied to select the aspect category and the aspect vector was concatenated with the word embedding, which was then further performed using CNN. Xu et al. [32] combined the CNN with the non-linear CRF to extract the aspect, which was then concatenated with the word embeddings and fed into another CNN to identify the sentiment polarity. Gu et al. [33] proposed a CNN with two levels that integrated the aspect mapping and sentiment classification. Compared with conventional ML approaches, this approach can lessen the feature engineering work and elapsed time [9]. It should be noticed that the performance of multitasking CNN does not necessarily outperform multitasking methods [19].

2.2.2 RNN and attention-based RNN

RNN has been applied for the ABSA and SBSA in the UGC. RNN models use a fixed-size vector to represent one sequence, which could be a sentence or a document, to feed each token into a recurrent unit. The main differences between CNN and RNN are: (1) the parameters of different layers in RNN are the same, making a fewer number of parameters required to be learned; (2) since the outputs from RNN relies on the prior steps, RNN can identify the context dependency and suitable for texts of different lengths [34–36].

However, the standard RNN has prominent shortcomings of gradient explosion and vanishing, causing difficulties to train and fine-tune the parameter during the process of prorogation [34]. LSTM and Gated Recurrent Unit (GRU) [37] have been proposed to tackle such issues. Also, Bi-directional RNN (Bi-RNN) models have been proposed in many studies [38, 39]. The principle behind Bi-RNN is the context-aware representation can be acquired by concatenating the backward and the forward vectors. Instead of the forward layer alone, a backward layer was combined to learn from both prior and future, enabling Bi-RNN to predict by using the following words. It has been proved that the Bi-RNN model achieved better results than LSTM in the highly skewed data in the task of aspect category detection [40]. Especially, Bi-directional GRU is capable of extracting aspects and identifying the sentiment in the meanwhile [23, 41] by using Bi-LSTM-CRF and CNN to extract the aspects in the sentence that has more than one sentiment targets.

Another drawback of RNN is that RNN encodes peripheral information, especially when it is fed with information-rich texts, which would further result in semantic mismatching problems. To tackle the issue, the attention mechanism is proposed to capture the weights from each lower level, which are further aggregated as the weighted vector for high-level representation [42]. In doing so, the attention mechanism can emphasize aspects and the sentiment in the sentence. Single attention-based LSTM with aspect embeddings [43], and position attention-based LSTM [44], syntactic-aware vectors [45] were used to capture the important aspects and the context words. The aspect and opinion terms can be extracted in the Coupled Multi-Layer Attention Model based on GRU [46] and the Bi-CNN with attention [47]. These frameworks require fewer engineering features compared with the use of CRF.

2.2.3 Memory network

The development of the deep memory network in ABSA was originated from the multi-hop attention mechanism that applies the exterior memory to compute the influence of context words on the given aspects [36]. A multi-hop attention mechanism was set over an external memory that can recognize the importance level of the context words and can infer the sentiment polarity based on the contexts. The tasks of aspect extraction and sentiment identification can be achieved simultaneously in the memory network in the model proposed by [13]. Li et al. [13] used the signals obtained in aspect extraction as the basis to predict the sentiment polarity, which would further be computed to identify the aspects.

Memory networks can tackle the problems that cannot be addressed by attention mechanism. To be specific, in certain sentences, the sentiment polarity is dependent on the aspects and cannot be inferred from the context alone. For example, "the price is high" and "the screen resolution is high". Both sentences contain the word "high". When "high" is related to "price", it refers to negative sentiment, while it represents positive sentiment when "high" is related to "screen resolution". Wang et al. [48] proposed a target-sensitive memory network proposed six techniques to design target-sensitive memory networks that can deal with the issues effectively.

2.3 Studies using DL methods in tourism and research gap

To obtain finer-grained sentiment of tourists' experiences in economy hotels in China, [11] used Word2Vec to obtain the word embeddings as the model input, and bidirectional LSTM with CRF model was used to train and predict the data. The whole model includes the text layer, POS layer, connection layer, and output layer, in which CRF was used for data output, reaching an accuracy of 84%. Chang et al. [49] applied GloVe to pre-train the word embedding. To improve the performance, feature vectors, like sentiment scores, temporal intervals, reviewer profiles, were added into CNN models. Their results proved that temporal intervals made a greater contribution than the sentiment score and review profile for the managers to respond to the reviews. Gao et al. [50] explored the model that built CNN on

LSTM and proved that the combined model outperformed the single CNN or LSTM model, with an improvement of 3.13% and 1.71% respectively.

To summarize, DL methods have been extensively used to perform ABSA. However, ABSA in the domain of tourism is little in the literature. Therefore, this study aimed at conducting ABSA using a dataset collected from TripAdvisor for predicting sentiments. Based on the literature review, it can be observed that RNN models especially attention-based RNN models achieved better performance than CNN models in terms of accuracy. Therefore, attention-based gated RNN models including LSTM and GRU were used in this study, which is summarized in the following section. Zhou et al. [14] conducted a series of ABSA on Semeval datasets [51, 52] using various DL methods. The experimental results confirmed that RNN with an attention-based mechanism obtained higher accuracies but relatively low precisions and recalls. This is because the Semeval datasets are naturally unbalanced datasets in which the fraction of positive sentiment samples is significantly higher than the fractions of neutral and negative sentiment samples, which indicates the importance of fractions of sentiment samples in the datasets. Inspired by ABSA on Semeval datasets, four datasets with different fractions of sentiment samples were resampled from the dataset of TripAdvisor hotel reviews to investigate the effect of sample imbalance on the model performance. Also, optimizers to minimize loss play a key role in model training. Therefore, three optimizers including the state-of-art optimizer were used in this study to compare their performance.

3. Research design and experiment

3.1 Corpora design

Based on the consideration and the purpose of the study, the corpora in this study will be completely in English and will include reviews collected from casino resorts in Macao. A self-designed tool programmed in Python was implemented to acquire all the URLs, which were first stored and further used as the initial page to crawl all the UGC that belongs to the hotel. The corpus includes 61544 reviews of 66 hotels. The length of the reviews varied greatly, with a maximum of 15 sentences, compared to the minimum of one sentence.

In terms of the size of the corpora that requires annotation, as there is no clear instruction regarding the size of the corpora, this study referred to Liu's work and SemEval's task. In machine learning based studies, it is reasonable to consider that the corpus that has 800–1000 aspects would be sufficient, while for deep-learning based approach, we think at least 5000 aspects in total would be acceptable. As the original data was annotated first to be further analyzed, 1% of the reviews were randomly sampled from the corpus. Therefore, 600 reviews that contain 5506 sentences were selected for ABSA in this study.

3.2 Annotation

Although previous works annotated the corpora and performed sentiment analysis, they did not reveal the annotation principles [51, 53] and the categories are rather coarse. For example, [53] used pre-defined categories to annotate the aspects of the restaurant. The categories involved "Food, Service, Price, Ambience, Anecdotes, and Miscellaneous", which did not annotate the aspects of finer levels. In addition, the reliability and validity of the annotation scheme have not been proved.

As the training of the models discussed above requires the annotation of domain-specific corpora, this study referred to [54]. The design of the annotation schema calls for the identification of aspect-sentiment pairs. Specifically, A is the collection of aspects a_j (with $j = 1, \dots, s$). Then, sentiment polarity p_k (with $k = 1, \dots, t$) should be added to each aspect in the form of a tuple (a_j, p_k).

To ensure the reliability and validity, Cohen's *kappa*, Krippendorff's *alpha*, and Inter-Annotator-Agreement (IAA) are introduced in this study, which are calculated by the agreement package in NLTK. Both indicators are used to measure (1) the agreement of the entire aspect-sentiment pair, (2) the agreement of each independent category.

3.3 Attention-based gated RNN

3.3.1 LSTM unit

The LSTM unit proposed by [25] overcomes the gradient vanishing or exploding issues in the standard RNN. The LSTM unit is consisted of forget, input, and output gates, as well as a cell memory state. The LSTM unit maintained a memory cell c_t at time t instead of the recurrent unit computing a weighted sum of the inputs and applying an activation function. Each LSTM unit can be computed as follows:

$$X = [h_{t-1} \ x_t] \tag{1}$$

$$f_t = \sigma\left(XW_f^T + b_f\right) \tag{2}$$

$$i_t = \sigma\left(XW_i^T + b_i\right) \tag{3}$$

$$o_t = \sigma\left(XW_o^T + b_o\right) \tag{4}$$

$$c_t = f_t \odot c_{t-1} + i_t \odot \tanh\left(XW_c^T + b_c\right) \tag{5}$$

$$h_t = o_t \odot \tanh(c_t) \tag{6}$$

where $W_f, W_i, W_o, W_c \in \mathbb{R}^{d \times 2d}$ are the weighted matrices, and $b_f, b_i, b_o, b_c \in \mathbb{R}^d$ are the bias vectors to be learned, parameterizing the transformation of three gates; d is the dimension of the word embedding; σ is the sigmoid activation function, and $\odot$ represents element-wise multiplication; x_t and h_t are the word embedding vectors and hidden layer at time$t$, respectively.

The forget gate decides the extent to which the existing memory is kept (Eq. (2)), while the extent to which the new memory is added to the memory cell is controlled by the input gate (Eq. (3)). The memory cell is updated by partially forgetting the existing memory and adding a new memory content (Eq. (5)). The output gate summarizes the memory content exposure in the unit (Eq. (4)). LSTM unit can decide whether to keep the existing memory with three gates. Intuitively, if the LSTM unit detects an important feature from an input sequence at an early stage, it easily carries this information (the existence of the feature) over a long distance, hence, capturing potential long-distance dependencies.

3.3.2 GRU

A Gated Recurrent Unit (GRU) that adaptively remembers and forgets was proposed by [37]. GRU has reset and update gates that modulate the flow of information inside the unit without having a memory cell compared with the LSTM unit. Each GRU can be computed as follows:

$$X = [h_{t-1} \quad x_t] \tag{7}$$

$$r_t = \sigma\left(XW_r^T + b_r\right) \tag{8}$$

$$z_t = \sigma\left(XW_z^T + b_z\right) \tag{9}$$

$$h_t = (1 - z_t) \odot h_{t-1} + z_t \odot \tanh\left(\left[r_{t \odot h_{t-1}} \quad x_t\right] W^T + b\right) \tag{10}$$

The reset gate filters the information from the previous hidden layer as a forget gate does in the LSTM unit (Eq. (8)), which effectively allows the irrelevant information to be dropped, thus, allowing a more compact representation. On the other hand, the update gate decides how much the GRU updates its information (Eq. (9)). This is similar to LSTM. However, the GRU does not have the mechanism to control the degree to which its state is exposed instead of fully exposing the state each time.

3.3.3 Attention mechanism

The standard LSTM and GRU cannot detect the important part for aspect-level sentiment classification. To address this issue, [43] proposed an attention mechanism that allows the model to capture the key part of a sentence when different aspects are concerned. The architecture of a gated RNN model considering the attention mechanism which can produce an attention weight vector α, and a weighted hidden representation r.

$$M = tanh\left(\begin{bmatrix} W_h H \\ W_v v_a \otimes e_N \end{bmatrix}\right) \tag{11}$$

$$\alpha = softmax(W_m M) \tag{12}$$

$$r = H\alpha^T \tag{13}$$

where $H \in \mathbb{R}^{d_h \times N}$ is the hidden matrix, d_h is the dimension of the hidden layer, N is the length of the given sentence; $v_a \in \mathbb{R}^{d_a}$ is the aspect embedding, and $e_N \in \mathbb{R}^N$ is a N-dimensional vector with an element of 1; $\otimes$ represents element-wise multiplication; $W_h \in \mathbb{R}^{d \times d}$, $W_v \in \mathbb{R}^{d_a \times d_a}$, $W_m \in \mathbb{R}^{d + d_a}$, and $\alpha \in \mathbb{R}^N$ are the parameters to be learned.

The feature representation of a sentence with an aspect h^* is given by:

$$h^* = tanh\left(W_p r + W_x h_N\right) \tag{14}$$

where $h^* \in \mathbb{R}^d$, W_p and $W_x \in \mathbb{R}^{d \times d}$ are the parameters to be learned.

To better take advantage of aspect information, aspect embedding is appended into each word embedding to allow its contribution to the attention weight. Therefore, the hidden layer can gather information from the aspect and the interdependence of words and aspects can be modeled when computing the attention weights.

4. Experiments and results

4.1 Annotation results

In the first trial, Cohen's *kappa* and Krippendorff's *alpha* are obtained at 0.80 and 0.78 respectively. Which are highly acceptable in the study since the scores

	Attribute	Polarity
First trial	0.86	0.88
Second trial	0.89	0.91

Table 1.
Cohen's kappa for categories of aspect and polarity.

measured the overall attribute and polarity. To identify the category that has the largest variation between two coders, Cohen's *kappa* for each label was calculated separately. Results (**Table 1**) indicated that Polarity had the highest agreement, while attribute showed lower agreement among two annotators. At the end of the first trial, both coders discussed the issues they encountered when they were annotating the corpus and make efforts to improve the preliminary annotation schema. The problems include dealing with the sentence that is difficult to assign the aspects.

Based on the revisions of the annotation schema, the coders conducted the second trial. With the revised annotation schema, the Cohen's *kappa* for the attribute and polarity is obtained at 0.89 and 0.91 respectively. In addition, Cohen's *kappa* and Krippendorff's *alpha* for the aspect-sentiment pair is computed by the end of the second trial, with 0.82 and 0.81 respectively, which indicated that the annotation schema in this study is valid.

4.2 Model training

The experiment was conducted on the dataset of TripAdvisor hotel reviews which contains 5506 sentences, where the numbers of positive, neutral, and negative sentiment samples are 3032, 2986, and 2725, respectively. Given a dataset, maximizing the predictive performance and training efficiency of a model requires finding the optimal network architecture and tuning hyper-parameters. In addition, the samples can significantly affect the performance of the model. To investigate the effect of sentiment sample fractions on the model performance, four sub-datasets with 4000 sentiment samples subjected to different sentiment fractions were resampled from the TripAdvisor hotel dataset as the train sets, one is a balanced dataset and three are unbalanced datasets that the sample fraction of sentiment positive, neutral, and negative dominated, respectively. In addition, it is observed that the average number of the aspects in a sentence is about 1.4, and the average length of the aspects in a sentence is about 8.0, which indicates that one sentence normally contains more than one aspect and the aspect averagely contains eight characters. The number of aspects in train and test sets is more than 850 and 320, respectively, which confirms the diversity of aspects in the dataset of TripAdvisor hotel reviews. For each train set, 20% of reviews were selected as the validation set.

Attention-based gated RNN models including LSTM and GRU were used for ABSA. Attention-based GRU/LSTM without and with aspect embedding were referred to as AT-GRU/AT-LSTM and ATAE-GRU/ATAE-LSTM, respectively. The details of the configurations and used hyper-parameters are summarized in **Table 2**. In the experiments, all word embeddings with the dimension of 300 were initialized by GloVe [17]. The word embeddings were pre-trained on an unlabeled corpus of which size is about 840 billion. The dimension of hidden layer vectors and aspect embedding are 300 and 100 respectively. The weight matrices are initialized with the uniform distribution U $(-0.1, 0.1)$, and the bias vectors are initialized to zero. The learning rate and mini-batch size are 0.001 and 16 respectively. The best

Configuration	Hyper-parameter
Word embedding	GloVe
Dimension of word embedding	300
Dimension of hidden layer	30
Dimension of aspect embedding	100
Initializer of weight matrices	Uniform distribution U(−0.1, 0.1)
Initializer of bias vectors	Zero
Optimizer	Search from {SGD, Adam, AdaBelief}
Number of epochs	Search from {100, 300, 500}
Dropout	0.5
Learning rate	0.001
Multi-batch size	16

Table 2.
Details of configurations and used hyper-parameters.

optimizer and number of epochs were obtained from {SGD, Adam, AdaBelief} and {100, 300, 500} respectively via grid search. The optimal parameters based on the best performance on the validation set were kept and the optimal model is used for evaluation in the test set.

The aim of the training is to minimize the cross-entropy error between the target sentiment distribution y and the predicted sentiment distribution $\hat{y}$. However, overfitting is a common issue during training. In order to avoid the over-fitting, regularization procedures including L2-regularization, early stopping as well as dropout were used in the experiment. L2-regularization adds "squared magnitude" of coefficient as a penalty term to the loss function.

$$loss = -\sum_i \sum_j y_i^j \, log \, \hat{y}_i^j + \lambda \|\theta\|^2 \tag{15}$$

where i is the index of review; j is the index of sentiment class, and the classification in this paper is three-way; λ is the L2-regularization term, which modified the learning rule to multiplicatively shrink the parameter set on each step before performing the usual gradient update; θ is the parameter set.

On the other hand, early stopping is a commonly used and effective way to avoid over-fitting. It reliably occurs that the training error decreases steadily over time, but validation set error begins to rise again. Therefore, early stopping terminates when no parameters have improved over the best-recorded validation error for a pre-specified number of iterations. Additionally, dropout is a simple way to prevent the neural network from overfitting, which refers to temporarily removing cells and their connections from a neural network [55]. In an RNN model, dropout can be implemented on input, output, and hidden layers. In this study, only the output layer with a dropout ratio of 0.5 was followed by a linear layer to transform the feature representation to the conditional probability distribution.

Optimizers are algorithms used to update the attributes of the neural network such as parameter set and learning rate to reduce the losses to provide the most accurate results possible. Three optimizers namely SGD [56], Adam [57], and AdaBelief [58] were used in the experiment to search for the best performance. The standard SGD uses a randomly selected batch of samples from the train set to compute derivate of loss, on which the update of the parameter set is dependent. The updates in the case of the standard SGD are much noisy because the derivative is not always toward minima. As result, the standard SGD may have a more time complexity to converge and get stuck at local minima. In order to overcome this issue, SGD with momentum is proposed by Polyak [56] (1964) to denoise derivative using the previous gradient information to the current update of the parameter set. Given a loss function $f(\theta)$ to be optimized, the SGD with momentum is given by:

$$v_{t+1} = \beta v_t - \alpha g_t \tag{16}$$

$$\theta_{t+1} = \theta_t + v_{t+1} \tag{17}$$

where $\alpha > 0$ is the learning rate; $\beta \in [0, 1]$ is the momentum coefficient, which decides the degree to which the previous gradient contributing to the updates of the parameter set, and $g_t = \nabla f(\theta_t)$ is the gradient at θ_t.

Both Adam and AdaBelief are adaptive learning rates optimizer. Adam records the first moment of gradient m_t which is similar to SGD with momentum and second moment of gradient v_t in the meanwhile. m_t and v_t are updated using the exponential moving average (EMA) of g_t and g_t^2, respectively:

$$m_{t+1} = \beta_1 m_t + (1 - \beta_1) g_t \tag{18}$$

$$v_{t+1} = \beta_2 v_t + (1 - \beta_2) g_t^2 \tag{19}$$

where β_1 and β_2 are exponential decay rates.

The second moment of gradient s_t in AdaBelief is updated using the EMA of $(g_t - m_t)^2$, which is easily modified from Adam without extra parameters:

$$s_{t+1} = \beta_2 s_t + (1 - \beta_2)(g_t - m_t)^2 \tag{20}$$

The update rules for parameter set using Adam and AdaBelief are given by Eqs. (23) and (24), respectively:

$$\theta_{t+1} = \theta_t - \frac{\alpha m_t}{\sqrt{v_t} + \varepsilon} \tag{21}$$

$$\theta_{t+1} = \theta_t - \frac{\alpha m_t}{\sqrt{s_t} + \varepsilon} \tag{22}$$

where ε is a small number, typically set as 10^{-8}.

Specifically, the update direction in Adam is $m_t/\sqrt{v_t}$, while the update direction in AdaBelief is $m_t/\sqrt{s_t}$. Intuitively, $1/\sqrt{s_t}$ is the "belief" in the observation, viewing m_t as the prediction of g_t, AdaBelief takes a large step when observation g_t is close to prediction m_t, and a small step when the observation greatly deviates from the prediction.

It is noted that the best models in the validation set were obtained by returning to the parameter set at the point in time with the lowest validation set error.

4.3 Results and analysis

As for the confusion matrix for a multi-class classification task, accuracy is the most basic evaluation measure of classification. The evaluation measure accuracy represents the proportion of the correct predictions of the trained model, and it can be calculated as:

$$Accuracy = \frac{\sum_{1}^{C} TP_i}{N} \tag{23}$$

where C is the number of classes (C equals to 3 in this study); N is the sample number of the test set; TP_i is the number of true predictions for the samples of the ith class, which is diagonally positioned in the confusion matrix. In addition to accuracy, classification effectiveness is usually evaluated in terms of macro precision and recall, which are aimed at a class with only local significance. As **Figure 1** illustrates, the class that is being measured is referred to as the positive class and the rest classes are uniformly referred to as the negative classes. The macro precision is the proportion of correct predictions among all predictions with the positive class, while macro recall is the proportion of correct predictions among all positive instances. The macro F1-score is the harmonic mean of macro precision and recall. The macro-average measures take evaluations of each class into consideration, which can be computed as:

$$MacroPrecision = \frac{1}{C}\sum_{i=1}^{C} \frac{TP_i}{TP_i + FP_i} \tag{24}$$

$$MacroRecall = \frac{1}{C}\sum_{i=1}^{C} \frac{TP_i}{TP_i + FN_i} \tag{25}$$

$$Macro - F1 = \frac{2 \times MacroPrecision \times MacroRecall}{MacroPrecision + MacroRecall} \tag{26}$$

where FP_i and FN_i are the number of false predictions for the positive and negative samples of the ith class, respectively.

This study computed accuracy (A), macro precision (P), macro recall (R), and macro F1-score (F) of AT-GRU, ATAE-GRU, AT-LSTM, and ATAE-LSTM trained with various optimizers and epochs. The results show: (1) Attention-based models (AT-GRU and AT-LSTM) performed better than attention-based models with

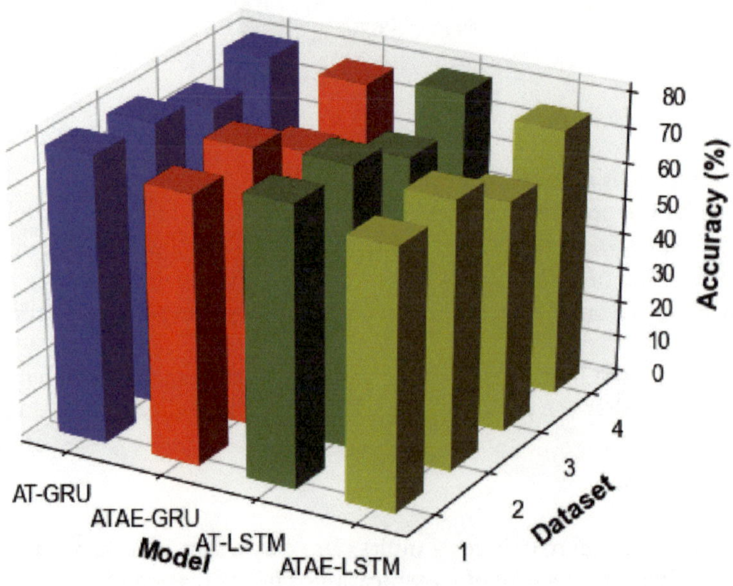

Figure 1.
Summary of model performance.

aspect embedding (ATAE-GRU and ATAE-LSTM). Taken Dataset 1 for example, the best accuracy in the test set using AT-GRU was 80.7%, while the best accuracy using ATAE-GRU was 75.3%; (2) Attention-based GRU performed better than attention-based LSTM. Taken AT-GRU and AT-LSTM for example, the accuracy and macro F1-score of AT-GRU for all datasets were higher than those of AT-LSTM; (3) The balanced dataset (Dataset 1) achieved the best predictive performance for all models. For the unbalanced datasets, the accuracy was exactly close to that of the balanced dataset. However, the macro precision, recall, and F1-score were significantly lower than those of the balanced dataset, which confirmed that the balanced dataset had the best generalization and stability in this study; (4) For Dataset 3 in which the neutral sentiment samples dominated, all of the models exhibited the worst predictive performance compared with other datasets. The candidate model for each dataset is illustrated in **Figure 1**. It is noted that the candidate model was selected according to accuracy. However, the model with a higher macro F1-score was selected as the candidate model instead when the accuracies of models were similar. Among 16 models, AT-GRU trained with the optimizer of AdaBelief and epoch of 300 in Dataset 1 achieved the highest accuracy of 80.7% and macro F1-score of 75.0% in the meanwhile. **Figure 2** illustrates the normalized confusion matrix of the best predictive model of which diagonal represented for the precisions. The precisions of positive and negative sentiment classification were about 20% higher than that of neutral sentiment classification, which confirmed that the need to boost the precision of neutral sentiment classification in order to globally improve the accuracy of the model in future work.

Early stopping was used in this research to avoid overfitting and save training time. **Figure 3** illustrates the learning history of AT-GRU using early stopping in four datasets, where the training stopped when the validation loss kept increasing for 5 epochs (i.e., "patience" equals to 5 in this study). For all datasets, the validation accuracy was exactly close to the training validation during the training procedure, which confirmed that early stopping was able to effectively avoid overfitting. Experimental results of A/P/R/F obtained based on training AT-GRU and AT-LSTM using early stopping. The accuracies obtained by AT-GRU and AT-LSTM were

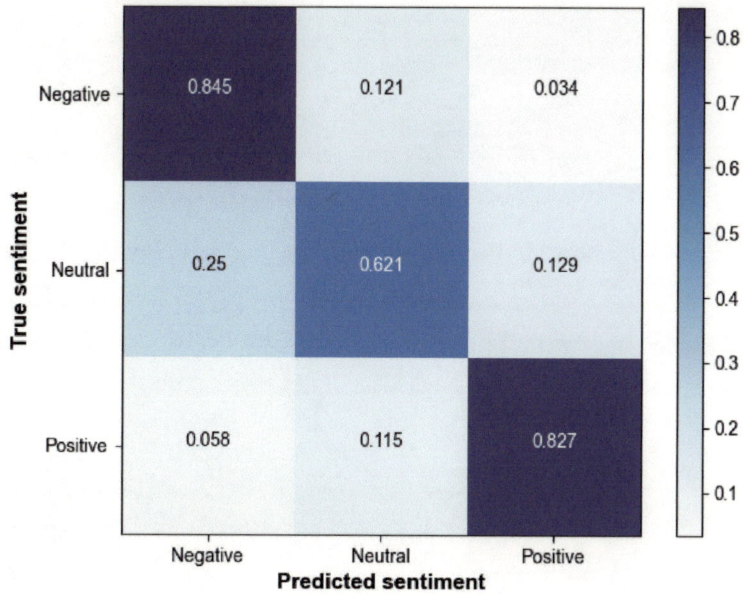

Figure 2.
Normalized confusion matrix of model with best predictive performance.

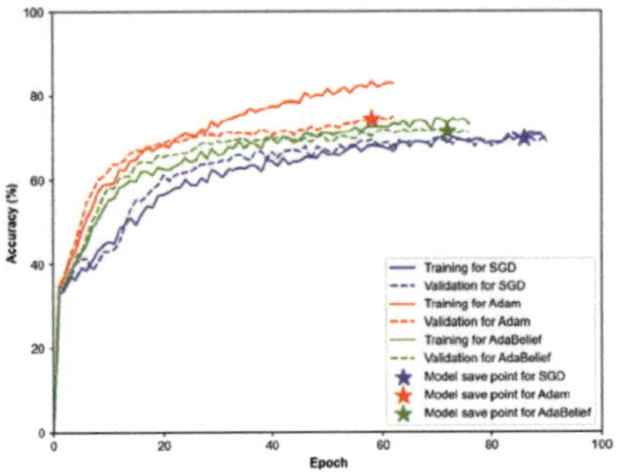

(a) Dataset 1

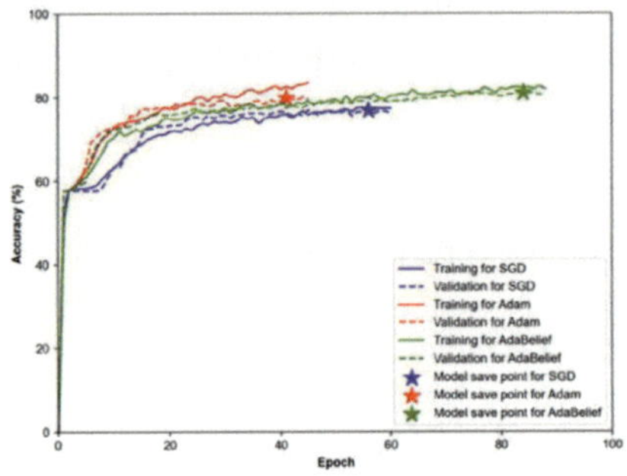

(b) Dataset 2

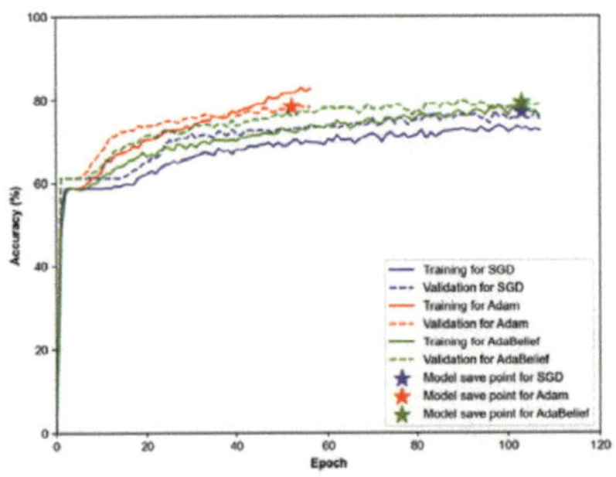

(c) Dataset 3

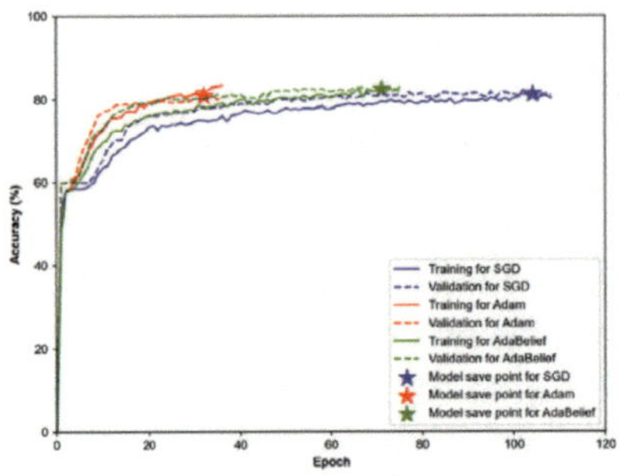

(d) Dataset 4

Figure 3.
Learning history of AT-GRU using early stopping.

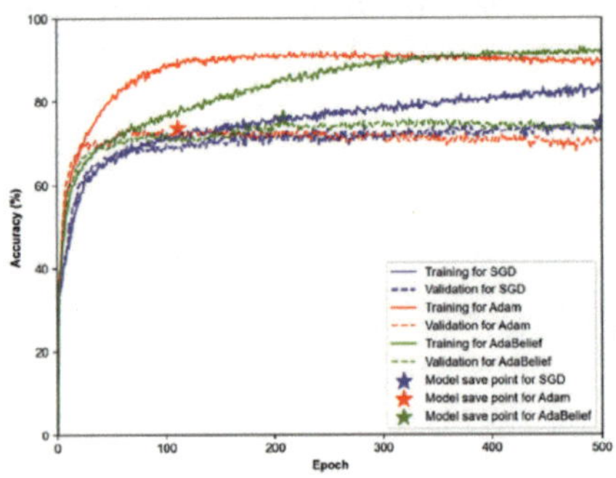

(a) Dataset 1

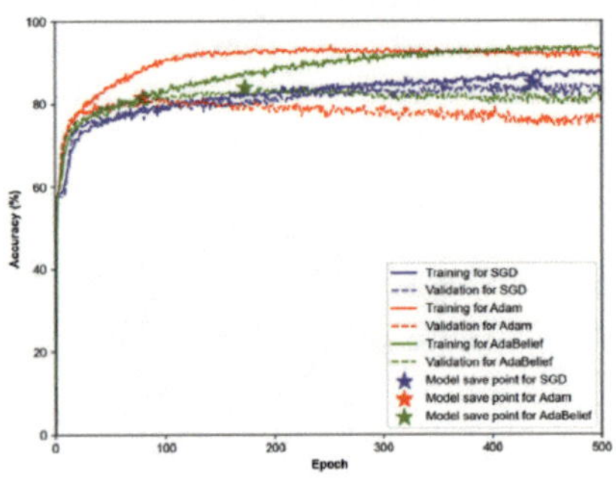

(b) Dataset 2

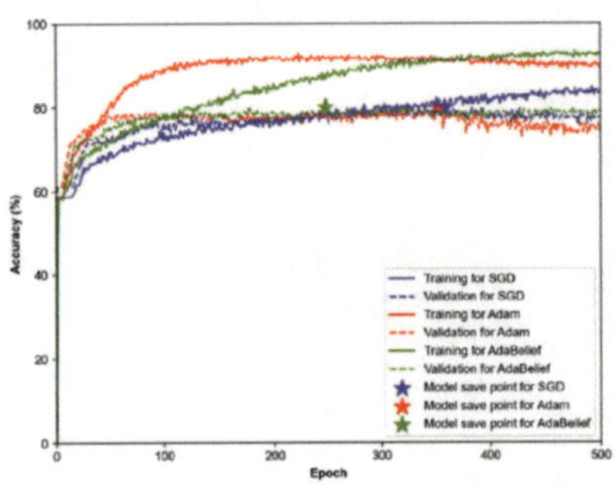

(c) Dataset 3

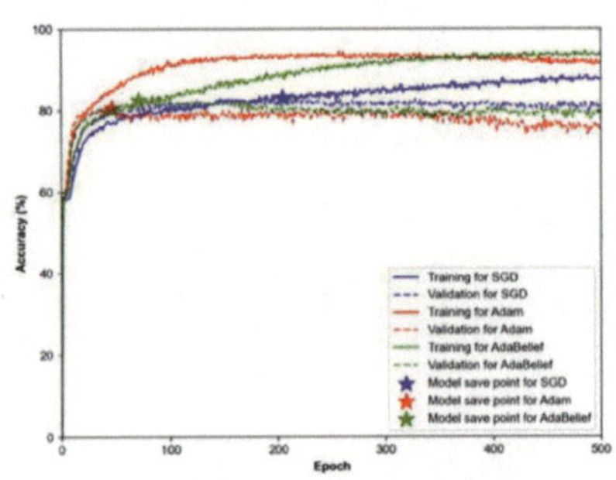

(d) Dataset 4

Figure 4.
Learning history of AT-GRU.

similar. For the balanced dataset, the accuracy and macro F1-score obtained by early stopping were significantly lower than that obtained by the corresponding model without early stopping. This is because the loss function probably found the local minima if the training stopped when the loss started to rise for 5 epochs. All of the optimizers used in this study were aimed at avoiding the loss function sticking at the local minima to find the global loss minima, therefore, using more epochs in the training was effective to obtain the best predictive performance model. On the other hand, for the unbalanced datasets, the accuracy and macro F1-score obtained by early stopping were similar to that obtained by the corresponding model without early stopping, which indicated that early stopping was effective to avoid overfitting as the loss converged fast in the unbalanced dataset. Although early stopping is a straightforward way of avoiding overfitting and improving training efficiency, the trade-off is that the model for test set possibly returns at the time point when reaching the local minima of loss function especially for the balanced dataset, and a new hyper-parameter of "patience" which is sensitive to the results is introduced.

Three optimizers were used in this study to find the best model. **Figure 4** illustrates the learning history of AT-GRU in four datasets. The gap between training and validation accuracy was the largest, which indicated that the worst generalization of Adam among three optimizers in this study although it converged quickly at the very beginning except for Dataset 3. Both SGD and AdaBelief can achieve good predictive performance with good generalization, however, AdaBelief converged faster than SGD, and the best results were achieved by AdaBelief.

5. Conclusions and future extensions

In this study, the hotel review dataset collected from TripAdvisor for aspect-level sentiment classification was first established. The dataset contains 5506 sentences in which the numbers of positive, neutral, and negative sentiment samples are 3032, 2986, and 2725, respectively. In order to study the effect of the fraction of sentiment samples on the model performance, four sub-datasets with a various fraction of sentiment samples were resampled from the TripAdvisor hotel review dataset as the train sets. The task in this study is to determine the aspect polarity of a given review with the corresponding aspects. To achieve a good predictive performance toward a multi-class classification task, attention-based GRU and LSTM (AT-GRU and AT-LSTM), as well as attention-based GRU and LSTM with aspect embedding (ATAE-GRU and ATAE-LSTM), were optimized with SGD, Adam, and AdaBelief and trained with epochs of 100, 300, and 500, respectively. Conclusions from these experiments are as follows:

1. AT-GRU and AT-LSTM performed better than ATAE-GRU and ATAE-LSTM. Taken the balanced dataset as an example, the best accuracy in the test set using AT-GRU was 80.7%, while the best accuracy using ATAE-GRU was 75.3%.

2. Attention-based GRU performed better than attention-based LSTM. Taken AT-GRU and AT-LSTM for example, the accuracy and macro F1-score of AT-GRU for all datasets were higher than those of AT-LSTM.

3. The balanced dataset achieved the best predictive performance. For the unbalanced datasets, the accuracy was exactly close to that of the balanced dataset, however, the macro precision, recall, and F1-score were significantly lower than those of the balanced dataset, which confirmed that the balanced dataset had the best generalization and stability in this study. For the dataset in

which the neutral sentiment samples dominated, all of the models exhibited the worst predictive performance.

4. For the balanced dataset, the accuracy and macro F1-score obtained by early stopping was significantly lower than that obtained by the corresponding model without early stopping. However, for the unbalanced datasets, the accuracy and macro F1-score obtained by early stopping were similar to that obtained by the corresponding model without early stopping, which indicated that early stopping was effective to avoid overfitting as the loss converged fast in the unbalanced datasets.

5. For optimizers, both SGD and AdaBelief can achieve good predictive performance with good generalization, however, AdaBelief converged faster than SGD, and the best results were achieved by AdaBelief.

This work includes the application of natural language processing technologies on the aspect-level sentiment analysis of the TripAdvisor hotel dataset, and there are still several extensions to be explored as follows:

1. Enlargement of the dataset. This study focused on the hotel in Macau, collecting 5506 reviews from TripAdvisor. To improve the model performance, hotels from other countries and regions can be collected into the dataset.

2. Improvement of model performance, especially for the predictive capacity of the neutral sentiment samples. The sentence and aspect embeddings were initialized with GloVe, and BERT which is popular in recent research can be used. In addition, although the attention mechanism was used in this study to improve model performance, the state-of-art self-attention mechanism such as multi-head attention can be used in the future to further refine the model.

3. Development of a mobile application. Once the model with stable performance is achieved, RNN algorithms can be integrated into a portable device such as a smartphone to help with real-time aspect-level sentiment analysis in tourism.

Author details

Weijun Li[1], Qun Yang[2] and Wencai Du[1*]

1 City University of Macau, Macao, SAR, China

2 Department of Civil and Environmental Engineering, The University of Auckland, Auckland, New Zealand

*Address all correspondence to: georgedu@cityu.mo

References

[1] Gretzel U, Yoo KH. Use and impact of online travel reviews. In: *Information and Communication Technologies in Tourism 2008*. Vienna: Springer Vienna; 2008. p. 35–46.

[2] Li H, Ye Q, Law R. Determinants of customer satisfaction in the hotel industry: An application of online review analysis. *Asia Pac J. Tour Res*. 2013;18(7):784–802.

[3] Choi S, Lehto XY, Morrison AM. Destination image representation on the web: Content analysis of Macau travel related websites. *Tour Manag*. 2007;28 (1):118–29.

[4] García-Pablos A, Cuadros M, Linaza MT. OpeNER: Open tools to perform natural language processing on accommodation reviews. In: *Information and Communication Technologies in Tourism 2015*. Cham: Springer International Publishing; 2015. p. 125–37.

[5] Kang H, Yoo SJ, Han D. Senti-lexicon and improved Naïve Bayes algorithms for sentiment analysis of restaurant reviews. *Expert Syst Appl*. 2012;39(5): 6000–10.

[6] Zheng W, Ye Q. Sentiment classification of Chinese traveler reviews by support vector machine algorithm. In: *2009 Third International Symposium on Intelligent Information Technology Application*. IEEE; 2009.

[7] Zhang Z, Ye Q, Zhang Z, Li Y. Sentiment classification of Internet restaurant reviews written in Cantonese. *Expert Syst Appl*. 2011;38(6): 7674–82.

[8] Liu B. Opinion Mining and Sentiment Analysis. In: Web Data Mining. Berlin, Heidelberg: Springer Berlin Heidelberg; 2011. p. 459–526.

[9] Do HH, Prasad PWC, Maag A, Alsadoon A. Deep learning for aspect-based sentiment analysis: A comparative review. *Expert Syst Appl*. 2019;118:272–99.

[10] Schouten K, Frasincar F. Survey on aspect-level sentiment analysis. *IEEE Trans Knowl Data Eng*. 2016;28(3): 813–30.

[11] Luo J, Huang S (sam), Wang R. A fine-grained sentiment analysis of online guest reviews of economy hotels in China. J *Hosp Mark Manag*. 2021;30 (1):71–95.

[12] Sheng F, Zhang Y, Shi C, Qiu M, Yao S. Xi'an tourism destination image analysis via deep learning. *J. Ambient Intell Humaniz Comput* [Internet]. 2020; Available from: http://dx.doi.org/ 10.1007/s12652-020-02344-w

[13] Li X, Bing L, Lam W, Shi B. Transformation networks for target-oriented sentiment classification. In: *Proceedings of the 56th Annual Meeting of the Association for Computational Linguistics* (Volume 1: Long Papers). Stroudsburg, PA, USA: Association for Computational Linguistics; 2018.

[14] Zhou J, Huang JX, Chen Q, Hu QV, Wang T, He L. Deep learning for aspect-level sentiment classification: Survey, vision, and challenges. *IEEE Access*. 2019;7:78454–83.

[15] Gu, S. Q., Zhang, L. P., Hou, Y. X., & Song, Y. A. A Position-aware Bidirectional Attention Network for Aspect-Level Sentiment Analysis. *Proceedings of the 27th International Conference on Computational Linguistics*. 2018;774–84.

[16] Mikolov T, Chen K, Corrado G, Dean J. Efficient estimation of word representations in vector space [Internet]. arXiv [cs.CL]. 2013. Available from: http://arxiv.org/abs/ 1301.3781

[17] Pennington J, Socher R, Manning C. Glove: Global vectors for word representation. In: *Proceedings of the 2014 Conference on Empirical Methods in Natural Language Processing (EMNLP)*. Stroudsburg, PA, USA: Association for Computational Linguistics; 2014.

[18] Poria S, Chaturvedi I, Cambria E, Bisio F. Sentic LDA: Improving on LDA with semantic similarity for aspect-based sentiment analysis. In: 2016 *International Joint Conference on Neural Networks (IJCNN)*. IEEE; 2016.

[19] Wu H, Gu Y, Sun S, Gu X. Aspect-based opinion summarization with convolutional neural networks. In: *2016 International Joint Conference on Neural Networks (IJCNN)*. IEEE; 2016.

[20] Liu P, Joty S, Meng H. Fine-grained opinion mining with recurrent neural networks and word embeddings. In: *Proceedings of the 2015 Conference on Empirical Methods in Natural Language Processing*. Stroudsburg, PA, USA: Association for Computational Linguistics; 2015.

[21] Bengio, Y., Schwenk, H., Senécal, J. S., Morin, F. M., & Gauvain, J. L. *Neural Probabilistic Language Models. Heidelberg*: Springer; 2006.

[22] Peters, M., Neumann, M., Iyyer, M., Gardner, M., Clark, C., Lee, K., & Zettlemoyer, L, editor. Deep contextualized word representations. In: *Proceedings of the 2018 Conference of the North American Chapter of the Association for Computational Linguistics: Human Language Technologies*. 2018. p. 2227–37.

[23] Devlin J, Chang M-W, Lee K, Toutanova K. BERT: Pre-training of deep bidirectional Transformers for language understanding [Internet]. arXiv [cs.CL]. 2018. Available from: http://arxiv.org/abs/1810.04805

[24] Feng J, Cai S, Ma X. Enhanced sentiment labeling and implicit aspect identification by integration of deep convolution neural network and sequential algorithm. *Cluster Comput*. 2019;22(S3):5839–57.

[25] Jebbara S, Cimiano P. Aspect-based sentiment analysis using a two-step neural network architecture. In: *Semantic Web Challenges*. Cham: Springer International Publishing; 2016. p. 153–67.

[26] MA, Y., Peng, H. Y., & Cambria, E. Targeted Aspect-Based Sentiment Analysis via Embedding Commonsense Knowledge into an Attentive LSTM. In: *The Thirty-Second AAAI Conference on Artificial Intelligence* (AAAI-18). 2018. p. 5876–83.

[27] Hochreiter S, Schmidhuber J. Long short-term memory. *Neural Comput*. 1997;9(8):1735–80.

[28] Goldberg Y. Neural network methods for natural language processing. *Synth lect hum lang technol*. 2017;10(1):1–309.

[29] Poria S, Cambria E, Gelbukh A, Bisio F, Hussain A. Sentiment data flow analysis by means of dynamic linguistic patterns. *IEEE Comput Intell Mag*. 2015; 10(4):26–36.

[30] Xue W, Li T. Aspect based sentiment analysis with gated convolutional networks. In: *Proceedings of the 56th Annual Meeting of the Association for Computational Linguistics* (Volume 1: Long Papers). Stroudsburg, PA, USA: Association for Computational Linguistics; 2018.

[31] Fan C, Gao Q, Du J, Gui L, Xu R, Wong K-F. Convolution-based memory network for aspect-based sentiment analysis. In: *The 41st International ACM SIGIR Conference on Research & Development in Information Retrieval -* SIGIR '18. New York, New York, USA: ACM Press; 2018.

[32] Xu L, Lin J, Wang L, Yin C, Wang J. Deep convolutional neural network-based approach for aspect-based sentiment analysis. In *Science & Engineering Research Support soCiety*; 2017.

[33] Gu X, Gu Y, Wu H. Cascaded convolutional neural networks for aspect-based opinion summary. *Neural Process Lett*. 2017;46(2):581–94.

[34] Goldberg Y. A primer on neural network models for natural language processing. *J. Artif Intell Res*. 2016;57: 345–420.

[35] Bengio Y. *Deep Learning*. London, England: MIT Press; 2016.

[36] Tang D, Qin B, Liu T. Aspect level sentiment classification with deep memory network. In: *Proceedings of the 2016 Conference on Empirical Methods in Natural Language Processing*. Stroudsburg, PA, USA: Association for Computational Linguistics; 2016.

[37] Cho K, van Merrienboer B, Gulcehre C, Bahdanau D, Bougares F, Schwenk H, et al. Learning phrase representations using RNN encoder–decoder for statistical machine translation. In: *Proceedings of the 2014 Conference on Empirical Methods in Natural Language Processing (EMNLP)*. Stroudsburg, PA, USA: Association for Computational Linguistics; 2014.

[38] Graves A. Supervised sequence labelling. In: *Studies in Computational Intelligence*. Berlin, Heidelberg: Springer Berlin Heidelberg; 2012. p. 5–13.

[39] Fan, Yuchen. Qian, Yao. Xie, Feng-Long. Soong, Frank K. TTS synthesis with bidirectional LSTM based recurrent neural networks. In: *INTERSPEECH*-2014. 2014. p. 1964–8.

[40] Chaudhuri A, Ghosh SK. Sentiment analysis of customer reviews using robust hierarchical bidirectional recurrent neural network. In: *Advances in Intelligent Systems and Computing*. Cham: Springer International Publishing; 2016. p. 249–61.

[41] Chen T, Xu R, He Y, Wang X. Improving sentiment analysis via sentence type classification using BiLSTM-CRF and CNN. *Expert Syst Appl*. 2017;72:221–30.

[42] Luong T, Pham H, Manning CD. Effective approaches to attention-based neural machine translation. In: *Proceedings of the 2015 Conference on Empirical Methods in Natural Language Processing*. Stroudsburg, PA, USA: Association for Computational Linguistics; 2015.

[43] Wang Y, Huang M, Zhu X, Zhao L. Attention-based LSTM for Aspect-level Sentiment Classification. In: *Proceedings of the 2016 Conference on Empirical Methods in Natural Language Processing*. Stroudsburg, PA, USA: Association for Computational Linguistics; 2016.

[44] Zeng J, Ma X, Zhou K. Enhancing attention-based LSTM with position context for aspect-level sentiment classification. *IEEE Access*. 2019;7: 20462–71.

[45] He, R., Lee, W. S., Ng, H. T., & Dahlmeier, D. Effective attention modeling for aspect-level sentiment classification. In: *Proceedings of the 27th International Conference on Computational Linguistics*. 2018. p. 1121–31.

[46] Wang, W., Pan, S. J., & Dahlmeier, D. Coupled Multi-Layer Attentions for Co-Extraction of Aspect and Opinion Terms. In: *Proceedings of the Thirty-First AAAI Conference on Artificial Intelligence* (AAAI-17). 2017.

[47] Cheng J, Zhao S, Zhang J, King I, Zhang X, Wang H. Aspect-level sentiment classification with HEAT (HiErarchical ATtention) network. In:

Proceedings of the 2017 ACM on Conference on Information and Knowledge Management - CIKM '17. New York, New York, USA: ACM Press; 2017.

[48] Wang S, Mazumder S, Liu B, Zhou M, Chang Y. Target-sensitive memory networks for aspect sentiment classification. In: *Proceedings of the 56th Annual Meeting of the Association for Computational Linguistics* (Volume 1: Long Papers). Stroudsburg, PA, USA: Association for Computational Linguistics; 2018.

[49] Chang Y-C, Ku C-H, Chen C-H. Using deep learning and visual analytics to explore hotel reviews and responses. *Tour Manag*. 2020;80(104129):104129.

[50] Gao J, Yao R, Lai H, Chang T-C. Sentiment analysis with CNNs built on LSTM on tourists' comments. In: *2019 IEEE Eurasia Conference on Biomedical Engineering, Healthcare and Sustainability* (ECBIOS). IEEE; 2019.

[51] Pontiki M, Galanis D, Pavlopoulos J, Papageorgiou H, Androutsopoulos I, Manandhar S. SemEval-2014 Task 4: Aspect Based Sentiment Analysis. In: *Proceedings of the 8th International Workshop on Semantic Evaluation* (SemEval 2014). Stroudsburg, PA, USA: Association for Computational Linguistics; 2014.

[52] Pontiki M, Galanis D, Papageorgiou H, Manandhar S, Androutsopoulos I. SemEval-2015 Task 12: Aspect Based Sentiment Analysis. In: *Proceedings of the 9th International Workshop on Semantic Evaluation* (SemEval 2015). Stroudsburg, PA, USA: Association for Computational Linguistics; 2015.

[53] Ganu, G., Elhadad, N., & Marian, A. (2009). Beyond the Stars: Improving Rating Predictions using Review Text Content [Internet]. Available from: Beyond the Stars: Improving Rating Predictions using Review Text Content. Twelfth International Workshop on the Web and Databases./http://spidr-ursa.rutgers.edu/resources/WebDB.pdf

[54] Moreno-Ortiz A, Salles-Bernal S, Orrequia-Barea A. Design and validation of annotation schemas for aspect-based sentiment analysis in the tourism sector. *Inf Technol Tour*. 2019;21 (4):535–57.

[55] Nitish Srivastava Geoffrey Hinton Alex Krizhevsky Ilya Sutskever Ruslan Salakhutdinov. Dropout: A Simple Way to Prevent Neural Networks from Overfitting. *Journal of Machine Learning Research*. 2014;15:1929–58.

[56] Polyak BT. Some methods of speeding up the convergence of iteration methods. *USSR Comput Math Math Phys*. 1964;4(5):1–17.

[57] Kingma DP, Ba J. Adam: A method for stochastic optimization [Internet]. arXiv [cs.LG]. 2014. Available from: http://arxiv.org/abs/1412.6980

[58] Zhuang J, Tang T, Ding Y, Tatikonda S, Dvornek N, Papademetris X, et al. AdaBelief optimizer: Adapting stepsizes by the belief in observed gradients [Internet]. arXiv [cs.LG]. 2020. Available from: http://arxiv.org/abs/2010.07468

Chapter 12

Data Mining Applied for Community Satisfaction Prediction of Rehabilitation and Reconstruction Project (Learn from Palu Disasters)

Andri Irfan Rifai

Abstract

Natural disasters can occur anytime and anywhere, especially in areas with high disaster risk. The earthquake that followed the tsunami and liquefaction in Palu, Indonesia, at the end of 2018 had caused tremendous damage. In recent years, rehabilitation and reconstruction projects have been implemented to restore the situation and accelerate economic growth. A study is needed to determine whether the rehabilitation and reconstruction that has been carried out for three years have met community satisfaction. The results of further analysis are expected to predict the level of community satisfaction for the implementation of rehabilitation and other reconstruction. The method used in this paper is predictive modeling using a data mining (DM) approach. Data were collected from all rehabilitation and reconstruction activities in Palu, Sigi, and Donggala with the scope of the earthquake, tsunami, and liquefaction disasters. The analysis results show that the Artificial Neural Network (ANN) and the support vector machine (SVM) with a DM approach can develop a community satisfaction prediction model to implement rehabilitation and reconstruction after the earthquake-tsunami and liquefaction disasters.

Keywords: Community Satisfaction, Data Mining, Disasters, Reconstruction, Rehabilitation

1. Introduction

The Palu earthquake, Indonesia, on September 28, 2018, caused severe damage with a reasonably broad impact. At the time of this writing, the atmosphere of grief and trauma of the people affected directly and indirectly began to disappear. The earthquake has a complete phenomenon in the movement of faults, tsunamis, landslides, and liquefaction events. Simultaneous liquefaction in several locations is unique in the world. This liquefaction phenomenon has received attention from the people in the world because the mudflow event during liquefaction has devastated infrastructure and housing on a massive scale [1].

Palu City and its surroundings based on topographic, geological, and seismological conditions can suffer damage due to earthquakes, including secondary disasters (tsunami, liquefaction, and cliff landslides). The earthquake in Palu on May 20, 1938, with a magnitude of 7.6 SR, was the previous incident with many fatalities. Studying, analyzing, and estimating all the supporting factors and the potential for disasters of such magnitude, the government needs to empower all components of society. The role of stakeholders in providing thoughts and recommendations is not accurate. Before and after an earthquake disaster occurs, they are better prepared psychologically and physically to reduce the impact of the disaster [2].

After a disaster with a significant impact, as mentioned above, various parties immediately carried out rehabilitation and reconstruction work, one of which was in transportation infrastructure. There are rehabilitation and reconstruction works on several roads, handling roads affected by liquefaction, including drainage systems, construction of retaining walls, construction of bridges, maintenance of bridges, and construction of access roads to permanent residences for disaster victims. According to its stages, the implementation of the rehabilitation and reconstruction was carried out, starting from recovery, trauma healing, permanent planning up to the overall reconstruction. The trauma healing stage is the starting point for the rehabilitation and reconstruction directly related to the community [3].

The implementation of rehabilitation and reconstruction due to natural disasters has not been completed yet. In early 2020 the Palu area could not avoid the non-natural disasters that plagued the world as a whole, namely the COVID-19 pandemic. This condition adds to the pressure to complete all stages of rehabilitation and reconstruction, especially work productivity which is directly impacted by restrictions on the labor movement. The decline in performance was mainly due to limited employee interactions with concerns and the potential risk of being exposed to the coronavirus. Covid-19 is transmitted by shedding droplets when an infected person coughs or exhales. Then, the released droplets will fall on nearby objects and surfaces, thereby polluting the surrounding environment [4].

Mitigation management and natural disaster recovery are an inseparable series of activities, starting from planning, mitigation, trauma healing, rehabilitation, and reconstruction, to socio-cultural recovery of the community. The speed and accuracy of planning play an essential role in achieving the success of post-disaster management. A thorough understanding and mapping are required in determining the plan that can be implemented appropriately in the field. Planning and implementation of work must consider the latest conditions taking into account the potential for recurring disasters. A thorough and well-targeted evaluation is required to ensure that the rehabilitation and reconstruction process runs according to the community's expectations. One of the evaluations that can be done is to measure community satisfaction at the job site. Because community satisfaction is one of the essential things in measuring the success of rehabilitation and reconstruction, the valuable experience from this disaster incident can be developed by a community satisfaction prediction model. The model that is built is expected to be an improvement step in the process of implementing rehabilitation and reconstruction in other activities.

2. Literature review

This section describes the literature review by conducting an integrated study of various information collected from library sources to provide a background for scientific development in rehabilitation and reconstruction. If necessary, comments and current knowledge trends will be included to show that the development of this knowledge can be included in the development of professionalism. In several sections, there is further information presented in different forms in implementing post-disaster rehabilitation and reconstruction. All information obtained from this literature review is used as a background to understand community satisfaction.

This paper will discuss about community satisfaction using a data mining approach. It is hoped that data mining can interpret and predict the data collected pre-during-post rehabilitation and reconstruction after the earthquake, tsunami, and liquefaction disaster. The use of data mining is believed to be able to provide a new approach in determining a better satisfaction level for the implementation of similar disaster management.

2.1 Disaster vulnerability

Apart from being famous for its wealth and natural beauty, Indonesia is also a country that is prone to disasters. This condition is because Indonesia is in a dynamic volcanic area and continental plates. This position also causes the shape of Indonesia's relief to varying widely, from mountains with steep slopes to gently sloping areas along very long coastlines, all of which are susceptible to landslide, flood, abrasion, and tsunami hazards. Various hydrometeorological conditions sometimes threaten flooding and landslides, hurricanes or tornadoes, drought-related forest fires, etc. Another threat is disasters caused by various technological failures.

The condition of Indonesia with a reasonably high risk of natural disasters such as Sulawesi Island is a complex area. The location of the Sulawesi is a meeting place for three large plates. The plate is the Indo-Australian Plate moving north, the Pacific Plate moving west, the Eurasian Plate moving south-southeast, and the smaller plate, the Eurasian plate, which moves south-southeast, and the smaller plate, the Philippine Plate. Sulawesi, a young island in Indonesia, is located where subduction and collisions are still active. Based on existing rock blocks, the island of Sulawesi can be divided into three parts of the geological area. The first is West Sulawesi, where tertiary deposits and magma rocks are the dominant parts. Second, Central and Southeast Sulawesi mainly consisting of rocks from the early Cretaceous era. Thrid, East Sulawesi ophiolitic nappe covered Mesozoic and Paleozoic era sedimentary rocks [5].

Palu City is one of the capital cities in Sulawesi, which has a high risk of disaster. Palu was also passed by a significant fault that divides the city firmly on the surface. This fault is often referred to as the Palu-Koro fault, originally called the Fossa Sarassina fault. All geologists and geophysicists who are familiar with the Palu-Koro fault agree that this fault is active. An active fault will experience an earthquake at the exact location of the period. Several studies show repeated earthquakes for hundreds and thousands of years [6]. These faults are thought to have caused the history of earthquakes in the area to be quite long. The history of earthquakes in central Sulawesi has been recorded since the 19th century. Several major earthquakes with a sufficiently large record were in 1968 with 6.7 SR, 1993 at 5.8 SR, and 2005 at 6.2 SR. Meanwhile, the tsunami occurred in 1927 in Palu Bay with a wave height of 15 m, 1968 in Malaga as high as 10 m, and 1996 in Simuntu Pangalaseang as high as 3.4 m [7].

This condition causes Palu's vulnerability to earthquakes to be very high. The studies about earthquake vulnerability by conducting a microtremor test in Palu City based on the earthquake's epicenter from the United States Geological Survey (USGS), magnitude 6.3, which occurred on January 23, 2005 [5]. Microtremor survey to estimate the distribution of solid earthquake vibrations. From the survey, the peak acceleration, velocity, and earthquake susceptibility index were obtained. From these observations, it can be concluded that Palu City has soil conditions with shear wave velocity Vs. < 300 m/s. The peak acceleration can reach more than 400-gal, resulting in significant damage to the building. From microtremor research, it is found that the vulnerability index in hilly areas is low and vice versa. The earthquake vulnerability index in the alluvium area is very high.

2.2 Rehabilitation and reconstruction

Rehabilitation is the repair and recovery of all public or community services to an adequate level in post-disaster areas. The main target of rehabilitation is to normalize or run fairly all aspects of government and community life in post-disaster areas. Rehabilitation is carried out by improving the environment in the disaster area, repairing public infrastructure and facilities, and providing assistance for community housing repairs. Rehabilitation activities also include socio-psychological recovery, health services, reconciliation and conflict resolution, socio-economic and cultural recovery, restoration of security and order. Furthermore, several other main activities that should not be neglected are restoring government functions and public services [8].

The implementation of rehabilitation includes physical repair activities and restoration of non-physical functions. Rehabilitation activities are carried out in areas affected by the disaster and other areas where it is possible to become target areas for rehabilitation activities. Rehabilitation activities must pay attention to building construction standards, social conditions, customs, culture, and economy. Repair of public infrastructure and facilities is an activity to repair public infrastructure and facilities to meet the transportation, smooth economic activities, and the socio-cultural life of the community [9].

Socio-economic and cultural recovery is part of the rehabilitation phase, aimed at helping communities affected by disasters to restore their social, economic, and cultural conditions to pre-disaster conditions. Social, economic, and cultural recovery activities are carried out by helping communities to revive and reactivate social, economic, and cultural activities through advocacy and counseling services, activity stimulant assistance, and training. This rehabilitation activity does not only concentrate on physical work but focuses more on social recovery. So the success of rehabilitation is not only measured by the recovery of physical conditions and infrastructure, but rather by the recovery of all community activities [10].

The next stage after or simultaneously with post-disaster management rehabilitation is reconstruction. In terms of handling reconstruction, a proper reconstruction process is needed, based on sound planning, so that it is right on target and orderly in the use of funds. It can increase community resilience to the threat of disasters in the future through disaster risk reduction efforts. A good post-disaster reconstruction process must recover community conditions, both physically, mentally, socially, and economically, and reduce vulnerability to disasters, not exacerbate existing vulnerability conditions that lead to disasters. For the reconstruction process to run well, it is necessary to involve non-governmental organizations and the general public [11]. The objective was to ensure the reconstruction process was planned on time, on quality, and budget, and following its objectives.

The reconstruction objective is to permanently rebuild part or all of the physical and non-physical facilities and infrastructure, along with the entire institutional and service system damaged by the disaster, so that conditions are restored. Their functions can run well, and the community can be better protected. From various catastrophic threats [12]. Resource mobilization, including human, equipment, material, and financial resources, is carried out by considering the available resources. Human resources who understand and have professional skills are indispensable in all post-disaster rehabilitation processes and activities. Resources in the form of equipment, materials, and funds are provided and ready to be allocated to support the rehabilitation and reconstruction process.

2.3 Community satisfaction

Monitoring of post-disaster rehabilitation and reconstruction is required to monitor disaster recovery processes and activities continuously. The steering committee and government elements carry out monitoring of rehabilitation and reconstruction activities. It may involve planning agencies at the national and regional levels as an overall ingredient in the implementation of rehabilitation [13]. Each rehabilitation program must meet specific achievement indicators, mainly so that each component of public infrastructure and facilities can function adequately again to support the resumption of the social and economic life of the people in the disaster area.

Disaster management activities are an inseparable series. One of the rehabilitation and rehabilitation phase implementations is an activity that must be linked to other stages. In this understanding, rehabilitation and reconstruction relate to the pre-disaster and emergency stages and trauma healing. The whole series of activities can be successful if each stage is carried out with strict monitoring and control. Therefore, disaster management should not be positioned as a goal but to achieve the efficiency and effectiveness of disaster management as a whole [14]. This condition is a necessity that obliges stakeholders to ensure that the planning, preparation, post-rehabilitation, and reconstruction stages are carried out under sound management principles.

In the rehabilitation and reconstruction phase, it is necessary to consider the available local resources to meet various implementation needs. Human resources who understand and have professional skills are indispensable in all post-disaster rehabilitation processes and activities. In addition, resources in equipment, materials, and funds are needed and are ready to be allocated to support the rehabilitation process [15]. Rehabilitation and reconstruction activities involving local communities can indirectly assist the community to revive social, economic, and cultural activities. It is hoped that the active involvement of the community in rehabilitation and reconstruction will make the community feel recognized as part of the community and ensure that community expectations are appropriately fulfilled.

The various steps taken during the rehabilitation and reconstruction phase must be ensured that they have met the community's needs or have not. In its stages, a community satisfaction survey is needed in connection with some of the above. This is a comprehensive measure of the level of community satisfaction with the quality of rehabilitation and reconstruction services provided by public service providers [16]. It is necessary to conduct a survey to determine the weaknesses of each indicator of public services. In addition, it can be used to determine the performance of the rehabilitation and reconstruction that has been carried out [17].

2.4 Data mining

Currently, soft computing methods are carried out by mimicking processes found in nature, such as the brain and natural selection [18]. Soft computing techniques make it possible to perform data processing to reduce uncertainty, imprecision, and ambiguity. In the mid-early 1960s, a new branch of computer science began to attract the attention of most scientists. This new branch, referred to as artificial intelligence (AI), can be defined as the study of how making computers drive the quality of people's work better. The AI approach encourages the development of soft computing in various fields, one of which is the development of data mining.

The development of the information technology industry is speedy, and knowledge in data collection is proliferating. Large databases are not a problem if they can take advantage of computer technology with various primary and supporting applications. All data collected and stored in a suitable database can be precious knowledge (for example, trend models, behavior models) that can support decision-making and optimize action [19]. Classical statistics have limitations for performing large amounts of data analysis or complex relationships between data variables. The solution for this problem and its limitations is to develop computer-based data analysis tools with more excellent capabilities and are automatic [20]. With the development of semi-automatic approaches in various fields of science, in recent decades, there has been an increase and across disciplines, such as AI, statistics, and information systems. This field is formally defined as knowledge discovery from the database (KDD). That in its development, KDD is increasingly known as DM [21].

One step in developing a community satisfaction prediction model in rehabilitation and reconstruction is processing the satisfaction data for each stage in a KDD process to form a DM prediction model. DM is a logical combination of data knowledge and statistical analysis developed in knowledge or a business process that uses statistical techniques, mathematics, artificial intelligence, artificial intelligence, and machine learning to extract and identify valuable information for related knowledge from large databases. The DM approach continues to be developed in various scientific fields. In recent times the use of DM for predicting social problems is increasing [22]. At the KDD stage, the DM algorithm has equipped a dataset used during the learning-phase, to be developed into a data-driven model. The model can be described as the relationship between input and output, which can provide helpful information.

Understanding and deepening the scientific field has an essential influence on the success of designing the DM algorithm. The database is only a meaningless set of data if an appropriate algorithm is not approached [23]. Furthermore, Fu also said that reviews carried out in the last few years show that DM's ability is growing in specific domains and depends on continuously developing specific algorithms. In simple cases, science can help identify the right features to model the data that underlie the compilation of scientific databases. Knowledge can also help design business goals that can be achieved using in-depth database analysis.

In this study, the database collects data on various satisfaction variables in the pre, during, and post-rehabilitation and reconstruction. Stages summarized in a post-disaster management system can be defined, and algorithms can be compiled to become real information support in improving mitigation management. The development of a system like this has a significant impact on the scientific development of disaster management, and even if the prediction accuracy is only a little, it is still better than random guessing. The availability of a complete database can provide a better and more reliable satisfaction prediction model [24].

3. Research method

In developing community satisfaction prediction models, complete information is needed about the characteristics of the type of work carried out. In general, community satisfaction at each stage is relatively easy to obtain if data is collected regularly and routinely. Community satisfaction is generally easy to compile and has several measurement methods to evaluate overall community satisfaction objectively. Meanwhile, data satisfaction that is outside the existing standard stages is a little more challenging to obtain and requires a long time. For example, data on community satisfaction pre-handling rehabilitation and reconstruction, compared to other stages, is more difficult to obtain. Existing data is more subjective, so that the quality of the data obtained depends on the ability of stakeholders to see and see analyze the conditions of these stages.

This section will describe the methods used to predict community satisfaction. This analysis is not mathematical, but it is carried out to obtain illustrations to show the argument that the proposed method is a more effective model. The community satisfaction prediction model is considered very important in completing a natural disaster management system. In addition, information related to the characteristics of community satisfaction includes pre, during, and post-rehabilitation and reconstruction, which are variables that are considered to have a significant influence on overall community satisfaction.

The community satisfaction model can be used in each stage, analyze disaster management, and determine the rehabilitation and reconstruction methods needed. Disaster management can analyze the existing conditions of the disaster management stages required to complete each disaster management step. This is linked to decision-making in management regarding the best and alternative methods for implementing post-disaster rehabilitation and reconstruction. In developing this model, researchers will use a DM-based community satisfaction prediction approach using data collected from the rehabilitation and reconstruction work locations in Palu, Sigi, and Donggala. Data is divided according to the handling area for calibration, learning, test, and validation purposes.

3.1 Model approach

This study will develop a community satisfaction prediction model with the DM approach without any restrictive assumptions by considering the input data sourced from the questionnaire results. The preparation of a community satisfaction prediction model with DM follows the following stages and processes. It was first cleaning and researching data that can be used in the deterioration model. The data cleaning process includes deleting inappropriate and irrelevant data from the database. This process can include writing errors, ensuring that the writing format remains consistent, and deleting records with incomplete data.

Second, check the data. The first step is to make a histogram or bar chart to determine the frequency of each variable. After that, the relationship of each data must be found. Knowing the distribution and correlation between existing variables helps researchers choose the proper form of data and be more efficient in evaluating the mode to be formed. In data checking, discrepancies and inaccuracies can be found so that further data cleaning is required. The level of correlation refers to the relationship between two variables. A high level of correlation indicates that the two variables are closely related, where if one of these variables changes, the other variables will also change proportionally. If the variables are continuous, these variables will form a line if drawn together. A low level of correlation indicates that the two variables change randomly and are not related. Most of the data fall between two extreme values. The correlation level test is shown through the correlation matrix.

Third, choosing the type of model. After considering each type of model previously studied (deterministic, probabilistic, and artificial intelligence). In this research, the development of the selected AI-based model. Developing a community satisfaction model is carried out through iteration stages by changing aspects of the model to form the best model based on the available data. Model development is done by adjusting aspects to the type of model and the available software. Several factors influence the shape of the model, among others, the basic equation, the variables used in the model, and the grouping of these variables into groups.

Fourth, look for parameter values. Determination of values and parameters is required in model development. In general, this step is completed using an optimized algorithm equation. However, for simple models (for example, a linear regression model using the least square method), this value can be manually optimized using a spreadsheet program. The *rminer* provides a complete menu option in determining the parameter value with the command:> contribution.

Finally, after the parameter values are obtained and the model has been formed, the model must be evaluated. The evaluation method will depend on the type of model selected. If, after evaluation, the model is not feasible, then the type of model must be reconsidered. If the type of model is still deemed inadequate, the form of the model must be changed and redeveloped. If the evaluation results conclude that the model type is unsuitable for the available data, then the model type must be reconsidered. There are several ways to evaluate statistical models. One of the initial actions that must be considered in evaluating a model is estimating parameter values. The parameter values must be reasonable and significant.

3.2 Model evaluation

By considering the classification or regression approach, other alternative evaluation steps can also be taken. The evaluation process is carried out for regression based on the difference between the observed value and the estimated value (error value). In general, the lower the error value, the better the community satisfaction prediction model, where the error value = 0 is the ideal value to be achieved.

In this study, three measurements were taken: the mean absolute deviation (MAD) root mean squared error (RMSE). Models with low MAD and RMSE values and R^2 values close to the unit value can be interpreted as models with a high level of prediction. RMSE is more sensitive to extreme values than MAD, and this is because RMSE uses the square value of the difference between the measurement results and the predicted model results. Compared to MAD, RMSE is more likely to produce a more significant error value in a model. Looking at the differences, measuring the error value through the two models will provide a different perspective on the proposed model to be used as a comparison.

Furthermore, different DM regression models can be easily compared by drawing a regression error characteristic (REC) graph, which depicts the tolerance for error values on the x-axis compared to the error tolerance percentage values estimated on the y-axis. The representation of the feasibility level of the model is also used in this study. All outputs are collected for evaluation. The integration of the R application with other reporting applications can be facilitated by compiling additional scripts.

3.3 R Tools

The satisfaction pattern through the community satisfaction prediction model is designed to be dynamic with various algorithm choices. The choice of the Multiple Regression (MR), ANN, and SVM algorithms is expected to provide various approaches to community satisfaction with the rehabilitation and reconstruction stages. The results of developing a community satisfaction model will be evaluated and adjusted throughout the disaster management stages until a model can translate the dynamics of existing data. The prediction model must be dynamic and respond to changing conditions [25].

Getting a fit model has carried out a whole iteration of all possible combinations between all variables. In this study, iterations were carried out with consideration of 25 variables and combination exploration. The model selection stage, especially during the feature selection stage, is only applied to the SVM algorithm. The advantage of this approach lies in the fact that the three SVM hyperparameters (c, γ, ϵ) can be set automatically and are urgently needed during the feature selection process.

During the learning phase (after selecting the input variables), the ANN algorithm in this study will use the overall multilayer perception relationship, with one hidden layer using H processing units, relationship predictions, and logistic activation functions 1 / (1 + e (−x)). The best value of H can be found by range {2, 4, ..., 10}, under the internal value (amount of training data used), around 5-fold cross-validation has been performed [26]. Based on tracing the built network, the value of H, which produces the smallest MAD value, has been selected, and ANN is retested using all training data. For the SVM algorithm, to reduce search space, this study uses the Gaussian kernel approach and the proposed heuristics approach to determine *complexity penalty parameter* = 3, and sizes for *incentive tube*, $\epsilon = \frac{\hat{\sigma}}{\sqrt{N}}$, where $\hat{\sigma} = \frac{1{,}5}{N} \cdot \sum_{i=1}^{N} (y_i - \hat{y}_i)^2$, y_i ais the amount of data used [27]. The most critical parameters in SVM are *kernel parameter* γ, used in the search scope $\{2^{-15}, 2^{-13}, ..., 2^{3}\}$, below the minimum 5*-fold cross-validation* [26].

Completing the modeling of the ANN and SVM algorithms, in this study, the MR model was tested as a comparison. The entire DM algorithm consisting of ANN, SVM, and MR is implemented with the R-Tool (R Development Core Team, 2009) and *rminer library* [28]. Furthermore, before fitting the ANN, SVM, and MR models, all data are tested with standard statistics, and then the output is tested for inverse transformation.

4. Experiment and discussion

As study material in this paper used data from the earthquake incident on September 28, 2018, in Palu, Sigi, and Donggala. This choice takes into account that the disaster has a reasonably broad impact on damage. In general, the damage can be divided into several phenomena. One of them is the damage caused by fault movements, fractures, and earthquake shocks. The fault movement is an offset where the left side moves north and the right side shifts to the south. The length of the most considerable shear on the right side is about 4 m, while the left side shifts to the north along 3 m. This shift is visible on the map visible on Google map. Of course, buildings that are traversed by faults will suffer significant damage and soil fractures, where fractures can be the impact of the movement of faults (or reactivated faults) with a smaller offset. Earthquake shocks are in the form of vibrations both horizontally and vertically. In general, in Palu City, the impact of damage due to shocks was not too much, except for buildings of low quality.

Therefore, is the phenomenon of damage due to the tsunami. The impact of a tsunami is the result of inundation (submerged buildings) and tsunami currents (speed or force acting to push or pull buildings). The impact of current velocity is mainly the scouring of the subgrade. If it is loose sand, the erosion rate is very high. Generally, buildings with shallow foundations fail because the scour reaches the base of the foundation. The buildings are relatively light, so they are easily carried away by the flow of water. Another damage is due to the tsunami and at the same time carrying debris to cars and ships, so collisions with these objects often result in heavy damage.

Lastly is the phenomenon of damage due to liquefaction. There are 4–5 locations that are pretty prominent and wide, namely in Balaroa, Petobo, Jono Oge, Lolu village (also in Jono Oge), and Sibalaya. Although some spots also occur liquefaction in the sand boil, it is not prominent and is not recorded. In addition, landslides in the sea can occur due to liquefaction. This kind of avalanche is induced by liquefaction. The landslides in Balaroa and Sibalaya were a phenomenon of liquefaction-induced landslides. It is possible that the submarine landslides that occurred in Palu Bay which caused the tsunami impact had the exact mechanism as in Sibalaya.

4.1 Community satisfaction prediction model

This section presents the modeling framework and procedures used to develop the ANN and SVM approach models. Similar to the traditional modeling process, where the goal is to estimate set coefficients in the form of a particular function. The main objective of the ANN model in this study is to obtain a set of matrices, which are abstract basic knowledge of the available data after going through the training loop. However, to use ANN in solving real-world problems, it is necessary to design a framework following the characteristics of a problem. The framework design aims to define the required ANN architecture and the relationships between the components in the framework. After completing the design framework, the next stage is to design the architecture of each ANN sub-model. The ANN architectural design process is a decision-making process, which includes determining the number of layers, the number of neurons in each layer, the variables entered into the input layer and the output layer. After completing the ANN architectural design, the design results need to be tested and validated.

In general, a neural network is made up of millions (even more) of the basic structures of interconnected and integrated neurons so that they can carry out activities regularly and continuously as needed. The imitation of a neuron in an artificial neural network structure is a processing element that can function as a neuron. The number of input signals is multiplied by the corresponding weight w. Then do the sum of all the results of the multiplication and the resulting output is passed into the activating function to get the degree of the output signal f (a, w). Although it is still far from perfect, the performance of this neuron clone is identical to that of the cell biology we know today. The collection of neurons is made into a network that functions as a computational tool. The number of neurons and the network structure for each problem solved is different.

Furthermore, this model was developed by activating the entire network in ANN. Activating an artificial neural network means activating every neuron used in that network. Many functions can be used as activators, such as goniometric and hyperbolic functions, step unit functions, impulses, sigmoid, etc. Of the several commonly used functions is the sigmoid function because it is considered closer to the human brain's performance. The algorithm activation process during iteration can be monitored, and its movement pattern can be seen.

In contrast to the neural network strategy, which seeks to find a hyperplane that separates classes, SVM tries to find the best hyperplane in the input space. The basic principle of SVM is a linear classifier. It is further developed to work on non-linear problems by incorporating the concept of a kernel trick in a high-dimensional workspace. This development encourages research in modeling to explore the potential capabilities of SVM theoretically and in terms of application. Currently, SVM has been successfully applied to real-world problems, and in general, provides a better solution than conventional methods.

4.2 Community satisfaction data

The model built is verified using data from questionnaire collection around the rehabilitation and reconstruction project. The questionnaire result dataset includes 625 results from 2 rehabilitation and reconstruction projects and 25 input parameters referred to as influencing parameters in an empirical study of community satisfaction. These parameters are given a sequence code based on the pre-during-post stage as input, as shown in **Table 1** below. All data obtained based on the level of importance and level of performance of each parameter asked the correspondent.

No	Code	Satisfaction Indicator
A. Before the rehabilitation and reconstruction		
1	A1	Information and socialization about reconstruction & rehabilitation
2	A2	The time the reconstruction program began
3	A3	Road & bridge damage identification process
4	A4	Participation in the reconstruction & rehabilitation process
5	A5	Collaboration between local communities in reconstruction & rehabilitation
6	A6	The wishes of the people are fulfilled by the reconstruction & rehabilitation
7	A7	Easy administration/disbursement process
8	A8	The role of government in the reconstruction process
B. During the rehabilitation and reconstruction		
9	B1	The role of the facilitator in the reconstruction & rehabilitation process
10	B2	Labor availability
11	B3	Work experience and skills
12	B4	Availability of material for reconstruction & rehabilitation
13	B5	Quality material available for reconstruction & rehabilitation
14	B6	Quality of road & bridge
15	B7	Community participation in the reconstruction & rehabilitation
C. After the rehabilitation and reconstruction		
16	C1	With the results of existing assistance
17	C2	The current state of the road & bridge is compared to the past
18	C3	The road & bridge become earthquake-resistant
19	C4	The comfort of road & bridge compared to before
20	C5	The quality of the road & bridge now compared to before
21	C6	The road & bridge was had been as a community wish
22	C7	Satisfaction with the current design
23	C8	The access road to residence compared to before the reconstruction & rehabilitation
24	C9	Current availability of street/environment lighting
Result		
25	CS	Community Satisfaction

Table 1.
Input code.

4.3 Stages of learning and modeling test

Forming a dataset is carried out to form three datasets that can be used immediately to learn, test, and validate. The database is divided into two datasets. The first set includes all the information. The dataset of both questionnaires was collected, which will be used for validation purposes. The entire dataset used for learning and test purposes is further divided into two subsets to obtain learning datasets. One set contains 80% of the data used for learning and 20% of the data used for testing. It is statistically independent data from the dataset used during learning and testing based on separating the dataset for the validation process. Therefore, verification of the DM model by using a separated dataset can be considered a control to check the performance of the DM model. The learning process is carried out with the number of epochs (10,000 times). The iteration process produces an ANN model that has an optimal weight between neurons.

After the learning phase is complete, the model development step is continued to the test stage to check the effectiveness of the learning process. The dataset used in the test stage becomes the DM input. The algorithm used in this stage uses a learning algorithm that has been recorded in the DM application when the learning process is running. The test process can calculate the error rate that occurs. If the error level of the test stage is still within an acceptable level, then the DM model is considered reasonable. A comparison of the model's accuracy is made by comparing the average MSE values during the test phase. Finally, the DM model with the lowest MSE error rate and the highest R^2 is selected. Finally, after the learning and test process is complete. Furthermore, the verification and validation of the model are carried out using the data that has been prepared with the prediction model of the community satisfaction learning and test results. Different dataset details were selected for model validation.

4.4 Model interpretation

In engineering science, apart from requiring a high level of accuracy, it also requires interpreting the modeling results. The ability to interpret DM is greatly influenced by the power of the data-driven model for this purpose. When the DM black box is implemented with ANN, SVM, and MR algorithms that involve complex mathematical expressions, the data-driven application procedure provided must translate the model. In this case, the results of the model interpretation are carried out to obtain a measurement of the input variables of the community satisfaction prediction model.

The first stage of model interpretation is to believe in the ability and accuracy of the model. The prediction model of community satisfaction using community satisfaction as the leading prediction parameter is first checked for modeling accuracy. There are several methods for evaluating predictive models, one of which uses the sum of absolute errors. The sum of the absolute errors often referred to as the absolute deviation of the average or MAD, is measuring forecasting accuracy by averaging the forecast errors using their absolute values. MAD is beneficial for analyzing and measuring the prediction error in the same unit of measure as the original data. In addition, the resulting process modeling criteria are stated in the RMSE, provided that the smaller the resulting RMSE (close to the value 0) will result in a better output prediction model.

This model is structured with a confidence level of 95% according to the t-student distribution. All DM models with ANN, SVM and MR algorithms are trained using 12 input variable attributes. **Figure 1** shows the predictive capacity of all training outcome models, comparing their performance in predicting the value of community satisfaction based on MAD, RMSE, and R^2. This table shows that the value of community satisfaction can be predicted accurately by each of the three DM models, especially by the ANN and SVM models.

Figure 1 below shows the standard error, and R^2 for each model developed. The DM model with the SVM algorithm has the smallest MAD value and RMSE value,

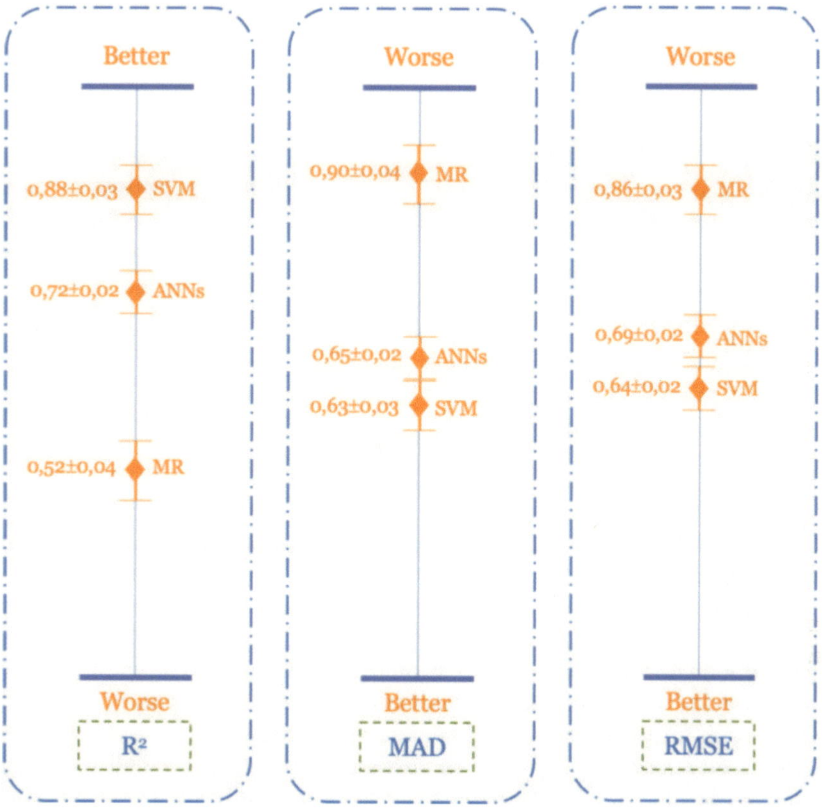

Figure 1.
Performance measured.

and the highest R^2 value. The prediction model with the ANN and SVM algorithms is acceptable and can be used in calculating community satisfaction predictions because it has R^2 close to 1. The following community satisfaction prediction model used in this study is the DM model with the SVM algorithm.

DM technique, also known as association rule mining, can find associative rules between a combination of items. Two parameters can determine the importance of an associative rule. The parameter is the percentage combination of these attributes in the database and confidence, namely the strength of the relationship between attributes in the associative rule. With the generate and test paradigm, the algorithm used in this study is making candidate combinations of attributes based on specific rules and then tested. Combining attributes that meet these requirements is called a frequent itemset, which is then used to create rules that meet the minimum confidence requirements.

By analyzing **Figure 2** (the scatterplot of the community satisfaction value prediction of the SVM algorithm with the questionnaire results), the variables that have been determined have a significant relationship with the change in the value of the questionnaire community satisfaction. **Figure 2a** shows the scatter-plots of learning results in the SVM model, and **Figure 2b** shows the results of the validation stages.

In the validation stage, the library feature *rminer* is used to describe and obtain the relative contribution value of each input value. The confirmed model has R^2, MAD, and RMSE values in the performance validation stage, such as **Figure 1**, with 20 runs performed, while the best hyperparameters to achieve a fit SVM model are used. $\epsilon = 0.07 \pm 0.01$ and $\gamma = 0.05 \pm 0.00$. Whereas the hyperparameters for ANN used $H = 3 \pm 1$.

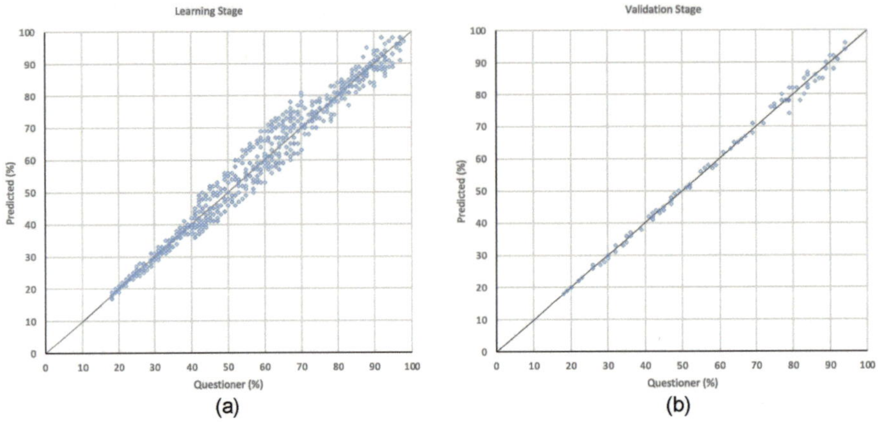

Figure 2.
Community satisfaction prediction outputs. a. Learning stage, b. validation stage.

Furthermore, the interpretation of the regression analysis used in DM is carried out. Package *rminer*, provides a graphical interpretation tool, namely: REC curve, error tolerance depicted on the x-axis, while the percentage value of road performance predictions is depicted on the y-axis. The resulting curve describes the level of error in the form cumulative distribution function (CDF). The error level defined as the difference between the predicted values of community satisfaction f(x) with community satisfaction actual on every coordinate (x, y). This approach is also a squared residual $(y-f(x))^2$ or absolute deviation $|y-f(x)|$ based on error metric mapping. **Figure 3** shown REC curve community satisfaction model with MR, ANN, dan SVM algorithm.

In **Figure 3** it can be analyzed that the REC curve describes the error tolerance on the x-axis and the level of accuracy of the regression function on the y-axis. The level of accuracy is defined as the percentage of modeling results that fit the specified tolerance. If the tolerance value is zero, only that value is considered to meet the model requirements. However, if you choose the maximum tolerance, other values can be used as reference for accuracy values. In the REC curve it is clear that the level of accuracy has a trade-off with tolerance. The greater the tolerance value given, the higher the accuracy value. Conceptually, the model with the lowest tolerance value with the highest accuracy is the model that has the best REC value.

The illustration of the REC curve depicts three different models. The curve shows that the SVM model has the highest accuracy value with the smallest tolerance value that moves consistently. This REC curve depicts the entire iteration process with 20 runs on the SVM model with hyperparameters as mentioned in the previous section. The shape of the REC curve can change shape when using different hyperparameters and the number of iteration runs is different.

4.5 Variable contribution

The DM model developed can assess each variable's contribution and attribute that becomes input data in the model. In this study, the variables or attributes consist of A1-C9. All attributes are then grouped into three dimensions pre, during, and post. A parameter vector in this DM model is chosen to explain that it is a variable function and not parameters as in the parametric approach. The only condition for a variance function is to be able to generate a non-negative definite variance matrix. Several methods can be used to estimate hyperparameter values. The value of θ can be predicted in this DM by using the cross-validation method. Hyperparameter used (H and γ) are H (2, 4, ..., 10) and γ (2–15, 2–13, ..., 23). This value produces

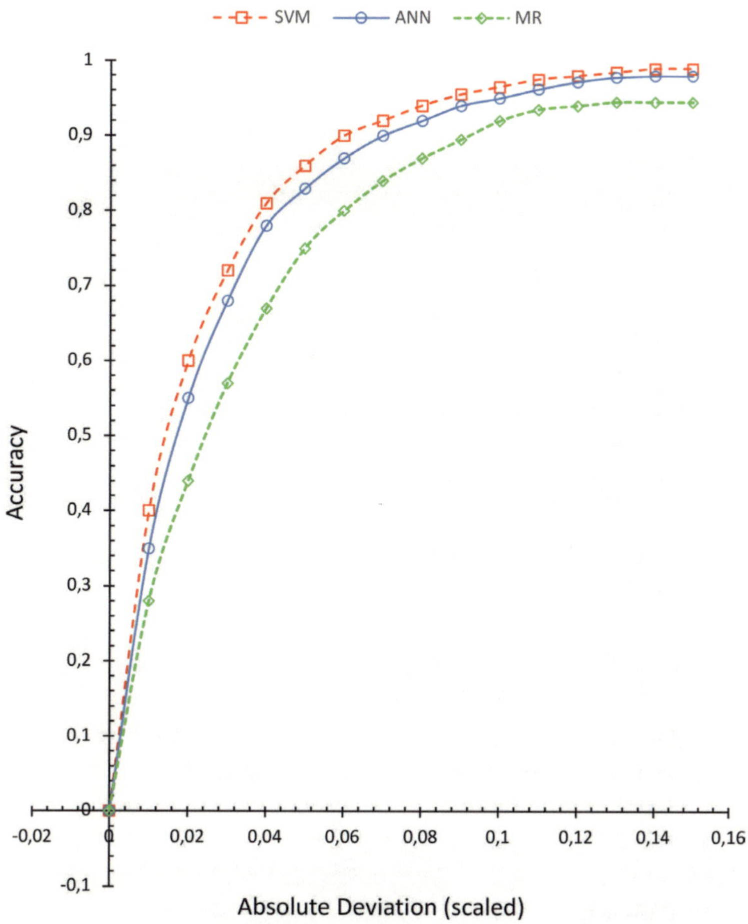

Figure 3.
The regression error characteristic curve.

the most precise model with optimal run time. For further model development, an approach can be used to try other hyperparameter values. The contribution of each attribute and dimension is of relative importance in composing the model.

The search results for the contribution value in DM can be simplified and displayed in **Figure 4**. This figure can display the relative importance on the x-axis for each attribute and dimension on the y-axis forming the community satisfaction prediction model with the DM model approach using the SVM, ANN, and MR algorithms.

Based on **Figure 4** below, each parameter has an almost even effect on community satisfaction in disaster management. When using a model that is considered the fittest, namely SVM, it can be seen that the most significant importance is the comfort of road and bridge compared to before (C4), and Collaboration between local communities in reconstruction and rehabilitation (A5). Therefore, the access road to residence compared to before the reconstruction and rehabilitation (C8), Participation in the reconstruction and rehabilitation process (A4), and Community Participation in the reconstruction and rehabilitation (B7). While pre-rehabilitation and reconstruction, the stage is the most critical dimension affecting community satisfaction.

The following model analysis is to compile an algorithm to select the main dimensions that affect the community satisfaction model and analyze the supporting variables that affect the community satisfaction prediction model that is not accommodated in this model. The results of VEC analysis illustrate the influence of the main attributes that move dynamically in the prediction model of community satisfaction

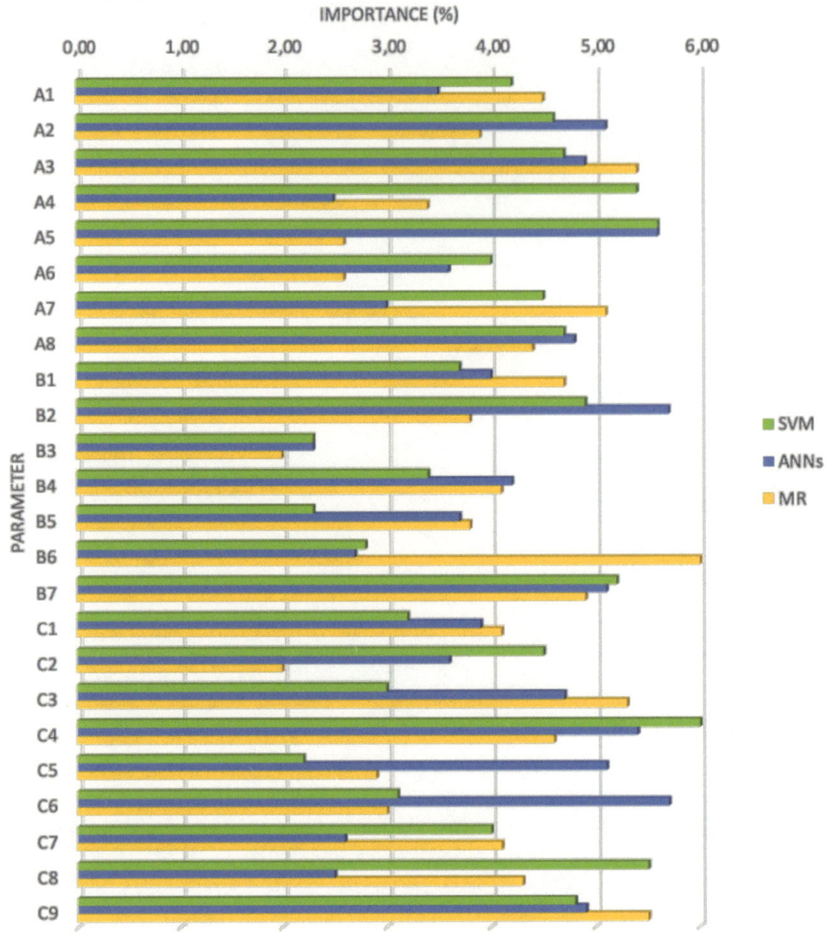

Figure 4.
Relative importance.

with this SVM model in the form of information and socialization about reconstruction and rehabilitation (A1), a pre-rehabilitation and reconstruction group. Decreased community satisfaction following the time of reconstruction program began (A2) and the role of the facilitator in the reconstruction and rehabilitation process (B1), and conversely, community satisfaction improved when performed the access road to residence compared to before the reconstruction and rehabilitation (C8).

5. Conclusion

The modeling process with the DM approach using the SVM, ANN, and MR algorithms produces a community satisfaction prediction model with a reasonably good model performance. The three model algorithms are compared with the questionnaire results. The REC curve shows the accuracy of each model used. Based on the resulting error matrix, it is believed that the SVM model is the best model to predict community satisfaction with a low iteration of 20 runs and has a good consistency. The most critical parameter in preparing the community satisfaction prediction model is the comfort of the road and bridge compared to before. Each attribute that affects the community satisfaction prediction model is successfully described with the algorithm of relative importance.

Acknowledgements

The authors are grateful to the editor and reviewers for their constructive comments on the earlier version of the paper. The Directorate General of Highway supported this research and liked to thank people for working at the Universitas Internasional Batam, Indonesia.

Author details

Andri Irfan Rifai
Universitas Internasional Batam, Batam, Indonesia

*Address all correspondence to: andri.irfan@uib.ac.id

References

[1] Tim Peneliti Unpar (2018). *Menyelisik Untaian Bencana Palu-Dongala (in Bahasa)*. Bandung: Universitas Katholik Parahyangan.

[2] Wu, X., Wang, Z., Gao, G., Guo, J., & Xue, P. (2020). Disaster probability, optimal government expenditure for disaster prevention and mitigation, and expected economic growth. *Science of the total environment, 709*, 135888.

[3] Meilianda, E., Munadi, K., Direzkia, Y., & Oktari, R. S. (2017). Assessment of post-tsunami disaster recovery of Banda Aceh city of Indonesia as window of opportunities for sustainable development. *IOP Conference Series: Earth and Environmental Science, Vol. 56, No. 1* (p. 012019). IOP Publishing.

[4] Jamaludin, S., Azmir, N. A., Ayob, A. F., & Zainal, N. (2020). COVID-19 exit strategy: Transitioning towards a new normal. *Annals of Medicine and Surgery*, 165-170.

[5] Thein, P. S., Pramumijoyo, S., Brotopuspito, K. S., Kiyono, J., Wilopo, W., Furukawa, A., & Putra, R. R. (2015). Estimation of S-wave velocity structure for sedimentary layered media using microtremor array measurements in Palu City, Indonesia. *Procedia Environmental Sciences, 28*, 595-605.

[6] Abdullah, A. I. (2020). A field survey for the rupture continuity of Palu-Koro fault after Donggala earthquake on September 28, 2018. *In Journal of Physics: Conference Series (Vol. 1434, No. 1)*, 012009.

[7] Widyaningrum, R. (2012). *Penyelidikan Geologi Teknik Potensi Liquefaksi Daerah Palu, Provinsi Sulawesi Tengah (in Bahasa)*. Bandung: Pusat Sumber Daya Air Tanag dan Geologi Lingkungan.

[8] Ahmad, M. I., & Ma, H. (2020). An investigation of the targeting and allocation of post-flood disaster aid for rehabilitation in Punjab, Pakistan. *International Journal of Disaster Risk Reduction, 44*, 101402.

[9] Nakamura, N., & Kanemasu, Y. (2020). Traditional knowledge, social capital, and community response to a disaster: resilience of remote communities in Fiji after a severe climatic event. *Regional Environmental Change, 20(1)*, 1-14.

[10] Islam, E., Abd Wahab, H., & Benson, O. G. (2020). Structural and operational factors as determinant of meaningful community participation in sustainable disaster recovery programs: The case of Bangladesh. *International Journal of Disaster Risk Reduction, 50*, 101710.

[11] Lu, Q., Zhong, D., & Zhang, Q. (2020). The evolving pattern of NGOs' participating in post-disaster community reconstruction in China: cases study on the 2008 Wenchuan earthquake and the 2013 Lushan earthquake. *Natural Hazards, 104(1)*, 167-184.

[12] Paudel, D., Rankin, K., & Le Billon, P. (2020). Lucrative Disaster: Financialization, Accumulation and Post-earthquake Reconstruction in Nepal. *Economic Geography, 96(2)*, 137-160.

[13] Daly, P., Mahdi, S., McCaughey, J., Mundzir, I., Halim, A., & Srimulyani, E. (2020). Rethinking relief, reconstruction, and development: Evaluating the effectiveness and sustainability of post-disaster livelihood aid. *International Journal of Disaster Risk Reduction, 49*, 101650.

[14] Yong, Z., Zhuang, L., Liu, Y., Deng, X., & Xu, D. (2020). Differences in the disaster-preparedness behaviors of the general public and professionals: evidence from Sichuan Province, China. *International journal of environmental research and public health, 17(14)*, 5254.

[15] Santha, S. D. (2018). Social interfaces in disaster situations: Analyzing rehabilitation and recovery processes among the fisherfolk of Tamil Nadu after the Tsunami in India. In *The Asian Tsunami and post-disaster aid* (pp. 65-78). Singapore: Springer.

[16] Sofyan, M. (2019). Community Satisfaction of the Urban Flood Control System Improvement Project (UFCSI). *Ilomata International Journal of Social Science, 1(1)*, 29-34.

[17] Ophiyandri, T., Hidayat, B., & Ghiffari, C. (2020). Community satisfaction levels on the housing reconstruction project after Mentawai tsunami in 2010-a case study at Sipora Island. *IOP Conference Series: Materials Science and Engineering Vol. 933, No. 1,* (p. 012041). IOP Publishing.

[18] Tinoco, J., Correia, A. G., & Cortez, P. (2014). Support vector machines applied to uniaxial compressive strength prediction of jet grouting columns. *Computers and Geotechnics 55*, 132-140.

[19] Liu, H., Motoda, H., Setiono, R., & Zhao, Z. (2010). Feature Selection: An Ever Evolving Frontier in Data Mining. *Proceedings of the Fourth Workshop on Feature Selection in Data Mining*, (pp. 4-13). Hyderabad, India.

[20] Rahman, F. A., Desa, M. I., Wibowo, A., & Haris, N. A. (2014). Knowledge Discovery Database (KDD)-Data Mining Application in Transportation. *Proceeding of the Electrical Engineering Computer Science and Informatics, 1(1),* (pp. 116-119.).

[21] Wang, X. Z. (2012). *Data mining and knowledge discovery for process monitoring and control.* Springer Science & Business Media.

[22] Suma, V., & Hills, S. M. (2020). Data Mining based Prediction of Demand in Indian Market for Refurbished Electronics. *Journal of Soft Computing Paradigm (JSCP), 2(02)*, 101-110.

[23] Fu, T. C. (2011). A review on time series data mining. *Engineering Applications of Artificial Intelligence, 24(1),* 164-181.

[24] Zhao, Y., Xu, X., & Wang, M. (2019). Predicting overall customer satisfaction: Big data evidence from hotel online textual reviews. *International Journal of Hospitality Management, 76,* 111-121.

[25] Rifai, A. I., Hadiwardoyo, S. P., Correia, A. G., Pereira, P., & Cortez, P. (2015). Data Mining Applied for The Prediction of Highway Roughness under Overloaded Traffic. *International Journal of Technology*.

[26] Hastie, R. T., Tibshirani, & Friedman, J. (2009). *The Elements of Statistical Learning: Data Mining, Inference, and Prediction.* Springer-Verlag New York, second edition.

[27] Cherkassky, V., & Ma, Y. (2004). Practical selection of svm parameters and noise estimation for svm regression. *Neural Networks, 17(1) ISSN 0893-6080.*, 113-126.

[28] Cortez, P. (2010). Data Mining with Neural Networks and Support Vector Machines Using the R/*rminer* Tool. *10th Industrial Conference on Data Mining (ICDM 2010)*, (p. Lecture Notes in Artificial Intelligence 6171). Advances in Data Mining.